A New Reality

HUMAN EVOLUTION FOR A SUSTAINABLE FUTURE

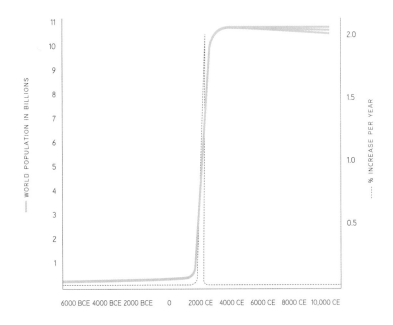

A New Reality

HUMAN EVOLUTION FOR A SUSTAINABLE FUTURE

Jonas Salk and Jonathan Salk

with David Dewane

City Point Press

CITY POINT PRESS
286 Curtis Avenue
Stratford CT 06615
www.citypointpress.com

Hardcover ISBN 978-1-947951-04-4
eBook ISBN978-1-947951-05-1

Cover and book design by Courtney Garvin

Manufactured in Canada

To the children of Epoch B,
   their children,
      and all the generations to follow.

The first edition of this book was made possible, in part, through the support of the United Nations Population Fund. The revised edition has received the support of private donors to the Arizona Community Foundation Collaborative Fund.

Any author's proceeds from sales of this book will be donated to the Jonas Salk Legacy Foundation and to the World Population and Human Values Fund of the Arizona Community Foundation.

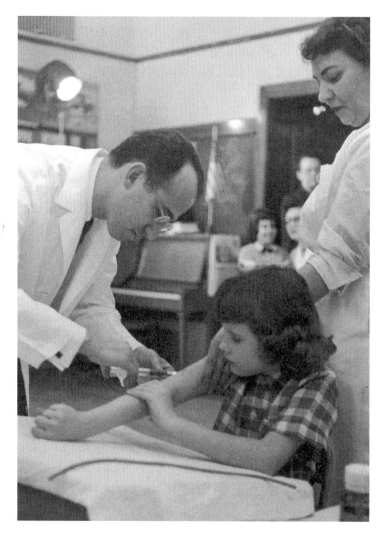

JONAS SALK, 1954

Jonas Salk's wish was that his ideas would continue to be disseminated so that, like a vaccine, they might have the most positive effect on the greatest number of people.

Jonas Salk, who died in 1995, was in the mid-1950s the developer of the first effective vaccine against poliomyelitis. He went on to found and help design the Salk Institute for Biological Studies in La Jolla, California, now a renowned center for basic biological research. What few people know is that in the last third of his life, he devoted much of his time and creative energy to the development of an evolutionary philosophy based on biological and natural principles. His wish was that these ideas would have the effect of giving people a scientific basis for hope and provide opportunities to enhance human well-being throughout the world.

Contents

FOREWORD

An inescapable reality is that there are now too many humans on Earth for us, or the planet, to handle well. Population growth, a burning topic some years ago, has somewhat faded from public prominence in recent years. This small but compelling book is a welcome and timely reminder of the issues of overpopulation, with a fresh look at the ways we can approach this reality.

The burgeoning world population of humans has commonly been framed in Malthusian terms, emphasizing Darwinism with its brutal selection of the "fittest" as an inevitable, and perhaps only, consequence of an overcrowded world. Jonas Salk took seriously, throughout his life, the overarching guideline called *Tikkun olam* (Hebrew: תיקון עולם) literally translated as "repair of the world," alternatively meaning "construction for eternity"). *Tikkun olam* is a concept in Judaism that has been taken to mean aspiration towards actions and behaviors that are constructive and beneficial. Jonas Salk's better-known life work, development of the polio vaccine that has been a life saver, literally, for millions around the globe, can be regarded as a remarkable exemplar of *Tikkun olam.* But Jonas Salk's legacy should rightly be broadened beyond even this, because the way he thought about the looming human population problem was to envisage a new era to which humanity could aspire.

In short, as is elegantly unfolded in this book (co-authored by Jonas Salk and his son Jonathan Salk), Jonas Salk envisioned that an inflection point in human population growth—a transition from exponential rapid growth, to slower and, eventually, zero population growth—would also usher in an inflection point in human social behaviors and mores, leading to a much more collaborative ethos and way of doing things. Rather than our latter-day humanity's central focus on competing in order to gain one's own individual betterment and achievements (defining what Salk dubs "Epoch A"), individuals would evolve toward ways more attuned to thinking beyond that, through expanding into wider and more generous frames of mind and spirit to encompass the needs, well-being, and attainment of many more, across more societies ("Epoch B"). Through use of simple diagrams and the building up of ideas, the book draws us gently but implacably into this vision.

Salk certainly is onto something here. While he was ahead of mainstream thinking at the time of the publication of the first edition of this book (1981), many of his ideas are already echoed much more commonly than they were then. As we look around us, in more recent years we see such trends at play in many arenas. Witness the evolving views of corporate leadership toward being more team- and participant-driven, rather than dictated by a sole top-dog figure. More and more academic learning and research are accomplished through fruitful interactions among multiple individuals, rather than solely through the lonely genius. And, *sine qua non*, we are realizing that to tackle shared world problems such as planetary climate challenges, individual, local, and national barriers get in the way. Thus, we will keep needing more worldwide, Paris Agreement–like, movements.

The extent of Jonas Salk's legacy deserves to be appreciated in full. As this book presages, his help in ameliorating humankind's scourges may yet turn out to be not confined to the near-eradication of polio. This elegant and hopeful book is small, but far from small in its vision and aspiration for humanity's betterment. We will all be better off if we listen to it and heed it.

Elizabeth H. Blackburn, PhD
February 2018

Preface

In this time of conflict, we are seeking a pathway into the future.

Half a century ago, thinking about the future of humanity, my father, Jonas Salk, had a realization. He looked at human population growth, which appeared to be increasing without limit, and reasoned that its growth would likely slow and reach a plateau. In doing so, it would form an S-shaped, or sigmoid, curve, similar to that of a population of fruit flies in a bottle.

From the image of that curve he developed a set of diagrams depicting our past and future and suggested that we are at an epochal transition in human history and human social evolution. He perceived that we are moving from an era dominated by limitless growth, competitive strategies, short-range thinking, and independence to one characterizedby awareness of limits, cooperation, long-range thinking, and interdependence.

The diagrams were first published in 1973 in his book *The Survival of the Wisest*. In 1981 he and I wrote *World Population and Human Values: A New Reality*, a short book expanding on those diagrams and ideas. We introduced images and a way of thinking that we hoped would provide a framework for understanding the nature of our time. In the decades since, I often noted that many of the changes we foresaw were coming to pass. Several years ago, a young architect and designer, David Dewane, came across the book and called me to see if I would like to revise and republish the original. I said yes, and the book you are holding is the result.

The first edition ended with these words:

We are on a frontier, but it is not territorial or technological; it is human and social. In this period of changing conditions and values, doubts arise as to our ability to cross this frontier and meet the demands of the future. We will, in the process of responding to forces and limits of nature, learn whether we have the capacity to meet this challenge. If we do, then we will emerge from the present period not merely as survivors, but as human beings in a new reality.

Using a series of figures opposite sparse text, the book presents graphs showing human population size over the long term, notes a remarkable change in its growth pattern, and suggests shifts that must take place if we are to survive and flourish in this transition.

The book's simplicity of design and concept belies a complexity of ideas. The discussion touches on population biology, demographic change, and socioeconomic conditions. It uses the analogy of evolutionary selection to look at shifts in human attitudes, values, and behavior. It addresses conflicts between differing value systems and their role in our evolutionary past and future. It considers the relationship between traditional and contemporary cultures. It concludes with an appreciation that the shift we describe involves all aspects of our existence—from the molecular, to that of the organism, to the societal, and to our relationship to the planet as a whole.

In the decades since the publication of the first edition, the world and our awareness of it have changed dramatically, yet the ideas and images in that slim volume seem even more relevant today. Population growth is slowing worldwide and will likely plateau by the end of this century. Limits in terms of energy and other resources have been encountered. Poverty and ill health are greatly reduced but remain endemic. Disturbingly, global climate change threatens our species and the nature of life on this planet. And most significantly, since the first appearance of the book, an entire generation has been born and is reaching adulthood in a period of decreasing population growth rates and awareness of limits—in a new reality.

In the course of change, there is conflict. The diagrams and concepts presented in the pages that follow address that conflict and suggest that the resolution of our current dilemma lies in the integration of opposing tendencies, one that will result in a synthesis that meets the needs and enhances the lives of all human beings. This point of view directs us to focus on the economic, political, and social changes we must make to adapt to a world with finite resources, more human beings than ever before, and a population that is at equilibrium or slowly declining.

While the figures and words contain a message of warning and danger, they also convey a message of hope. They point to our shared responsibility for the planet and our species, and they also provide, especially for those of younger generations, images and a vision that can be used to shape the future.

The current edition differs considerably from the original in both text and design. I have done my best to stay with the tone and style of the original document, but I confess that I have often missed my father's guiding hand. There was a quality that he brought to his writing—a complex focus and intensity that were uniquely his.

The book represents a type of intuitive thinking that was typical of my father, who often made creative leaps and saw connections that others might not see. As such, it is not intended to be an exhaustive synthesis of academic research, nor does it attempt to be a prescriptive document, laying out specific plans of action. It does aspire, however, to shape our thinking and our perception. It is also, I hope, a bit of an elegy to my father's work, vision, and dedication to helping humanity.

My father's desire was that the ideas in his writing be highly accessible so that they might have a positive effect on people and on the course of human evolution. Typical of him, his aim was grand. Also typical of him, it was not beyond the bounds of reality. He wanted to put forth words and images that would affect people in a way similar to a vaccine, inoculating them with hope; immunizing them against stasis, rigidity, and despair; and allowing all human beings to live fuller, more creative, and more productive lives.

With humble respect for that grand idea, we present this revised and redesigned edition, now titled *A New Reality: Human Evolution for a Sustainable Future*

Jonathan Salk
Los Angeles, 2018

Even though the discussion here suggests that, in the long term, the course of epochal change is predetermined, it is not.

It is under our influence.

PART ONE

Sigmoid Curves

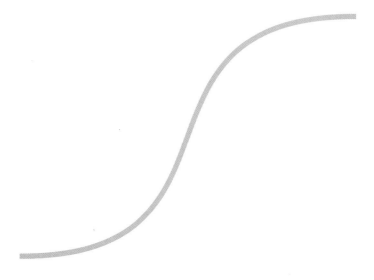

In this essay, the sigmoid curve will be used as a "thinking tool" and as a symbol. Its shape reflects a pattern that applies to growth in living systems and reflects the transformational character of change in our time.

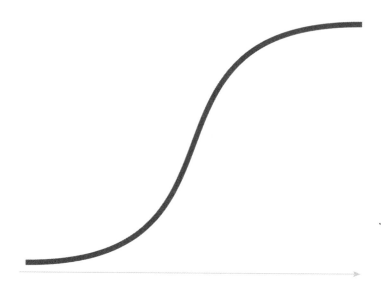

In this figure, and in those that follow, the horizontal axis represents time...

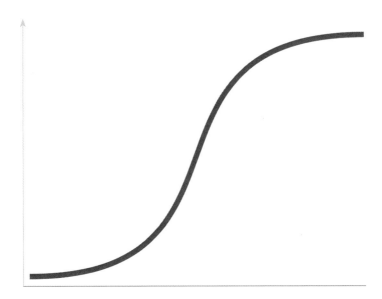

...and the vertical axis represents number.

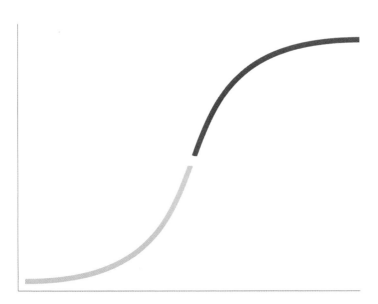

In the first, upturned portion of
the curve, population growth follows
a pattern of acceleration;

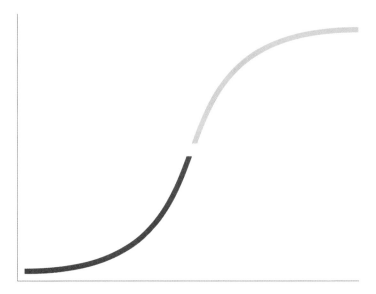

in the second part, growth decelerates
and a plateau is reached.

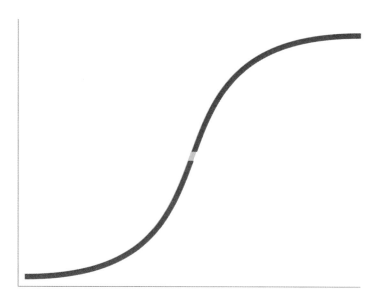

The gap in the curve emphasizes the *point of inflection* — the point of change from accelerating growth to decelerating growth.

From the beginning of the Common Era, the size of the human population grew gradually for about 16 centuries and then with increasing speed through the 19th century. The gradual but progressive acceleration was followed by a sudden steep rise in the 20th century—a consequence of the scientific, technological, industrial, and agricultural revolutions, which have had the effect of making it possible to sustain a human population far larger than ever before.

CE	200	400	600	800	1000

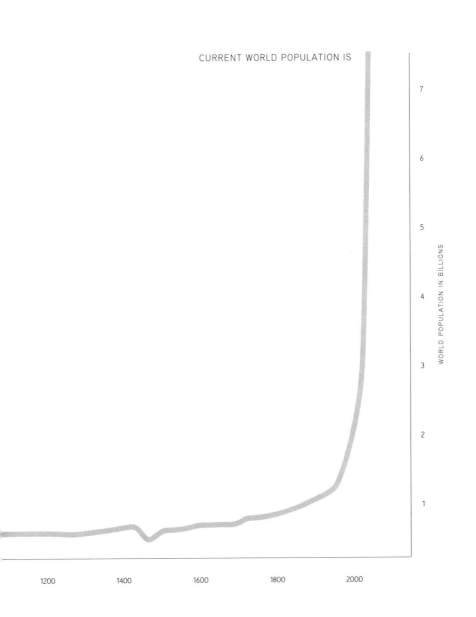

CURRENT WORLD POPULATION IS

WORLD POPULATION IN BILLIONS

7

6

5

4

3

2

1

1200   1400   1600   1800   2000

The sharp increase in the size of
human population in recent times
raises reasonable questions:

Will the curve continue to rise at
its present rate?

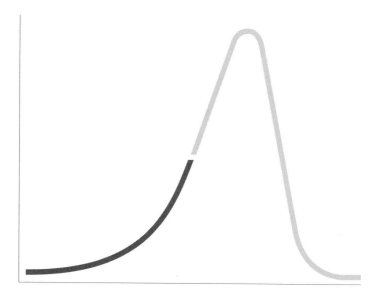

Will it crash?

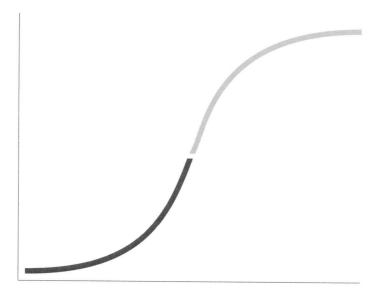

Or will it bend and assume
a sigmoid shape?

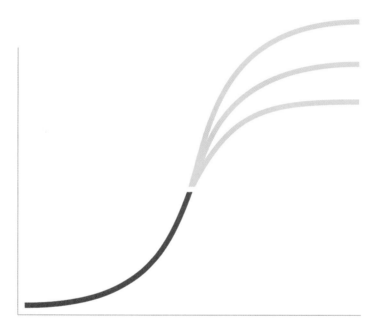

If the curve will inflect, when might this occur and at what level will population size plateau?

That an inflection can be anticipated is suggested by the figures in the pages that follow. In part 2, we will see that the inflection of the human population growth curve has already occurred; the height of the plateau is still uncertain, however, and is subject to our influence.

The S-shaped growth curve is seen in many living systems. We will present a few examples before focusing on human population growth.

Fruit Fly
DROSOPHILA MELANOGASTER

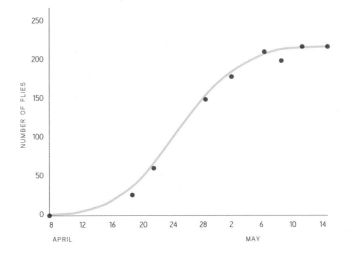

This figure is a plot of the growth of a population of fruit flies in a laboratory experiment. A small number of flies were introduced into a chamber of fixed size, and the increase in population was observed over a period of approximately five weeks. The population grew slowly at first and then more rapidly as the number of flies increased. After about two and a half weeks, however, growth began to slow. Over the next two and a half weeks, the number of flies that were being born approached the number that were dying, and the curve reached a plateau.

Yeast Cells
SACCHAROMYCES CEREVISIAE

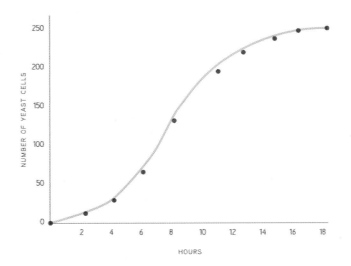

The sigmoid pattern is also observed in the growth of a population of microorganisms. This figure indicates that a population of yeast cells, over a period of 18 hours, follows a similar pattern of increase in the growth rate followed by decrease and the attainment of a plateau.

This and the preceding figure indicate that under the conditions of these two studies, population growth slows and reaches a plateau. The following figure indicates what happens when conditions are changed.

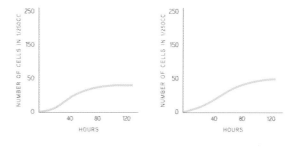

CONTROL GROUP   ACIDITY HELD CONSTANT

The effect of external influences on growth is shown here in observations made of yeast cultures maintained over periods extending up to 120 hours. The growth curve of the control culture (top left) reveals a sigmoid pattern.

By neutralizing the acid produced in the course of growth—a process that ameliorates the negative effects of waste produced in the environment (a process equivalent to improved sanitation)—a noticeably higher plateau was reached, as seen in the curve at top right.

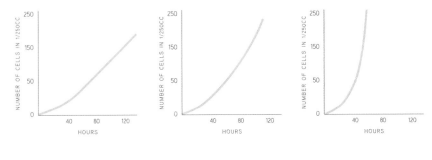

MEDIUM CHANGED
EVERY 24 HRS.

MEDIUM CHANGED
EVERY 12 HRS.

MEDIUM CHANGED
EVERY 3 HRS.

These three curves show the effect that increase in
food supply and waste removal produced by changing
the culture medium every 24 hours, every 12 hours,
or every 3 hours. We see a steepening of the curve
under these conditions. To sustain population growth
at these levels would require an uninterrupted input
of food and the simultaneous elimination of the waste
products of metabolism. Failure of either would result
in a catastrophic collapse in population size.

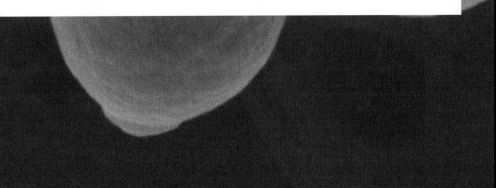

Sheep
OVIS ARIES

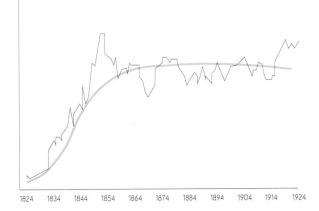

In this study of a population of sheep on the island of Tasmania, in southern Australia, we see that after a period of exponential growth, a plateau was reached approximately 30 years after the introduction of the animals onto the island and that this plateau continued for at least 70 years, the full period of observation.

Each of the preceding sigmoid curves exhibits different rates of change, some counted in minutes, some in hours, some in days, months, or years. Various populations react in different time frames depending on the nature of the organisms and on the conditions in which they live.

Through examples of growth of populations in closed systems, the sigmoid curve was introduced.

After a period of slow growth, human population size has increased sharply in recent centuries. Except for the sun as a constant source of energy, the earth can be seen as a closed system, and by inference from the examples given, we can expect the human population growth curve to follow a sigmoid pattern. The level of the plateau is still uncertain, however, and is subject to human influence.

In the years to come, we face the challenge of understanding and facilitating a slowing of human population growth and, ultimately, of adapting to conditions associated with a relatively constant population size at a level far beyond anything we have previously experienced.

PART TWO

World Population Trends

For 200,000 years before the advent of agriculture, human population increased very slowly. Agriculture emerged 10,000 to 15,000 years ago, making more food and energy available to support greater numbers of human beings. A pattern of gradual increase then continued throughout the agrarian period. In the last two centuries, scientific, technological, industrial, and agricultural developments have reduced mortality and made it possible to support and feed far larger numbers of people, resulting in the recent sharp rise in population.

The factors involved in human population growth are far more complex than those affecting the populations seen in part 1. The picture is complicated by family, cultural, sociopolitical, economic, and technological factors. Marked differences between patterns of population growth in the more and less developed regions of the world are also significant. Nevertheless, in part 2 we will see that, taken globally, a similar pattern is emerging.

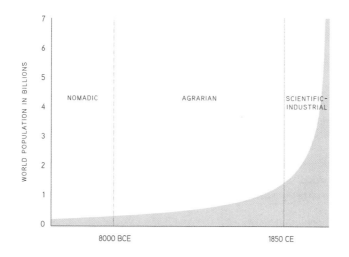

This graph shows the growth of human population over the last 2,000 years. It took from the beginning of our species, 200,000 years ago, until 1804 for the population to reach 1 billion. The addition of the next billion took only 123 years. Since then, each billion has been added over a period shorter than the previous—the population growing at a faster and faster rate.

The time taken to add a billion people reached its low point, 12 years, with the increase from 5 to 6 billion in 1998. The next shift, from 6 to 7 billion, also took 12 years. It is projected that it will take 13 years to reach 8 billion.

While population increased faster and faster into the last part of the 20th century, it is clear that growth is now slowing. Acceleration has changed to deceleration in overall population growth, and as we will see, the point of inflection in the sigmoid curve has been passed.

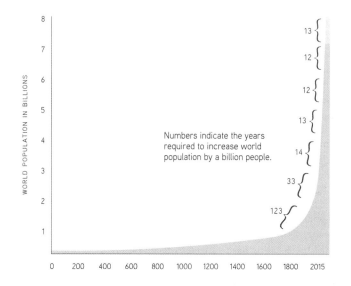

Numbers indicate the years required to increase world population by a billion people.

This figure narrows the focus, showing population growth from 1890 to the present. We see that, when looking at a shorter time period, the growth curve resembles the first part of the curves seen in part 1.

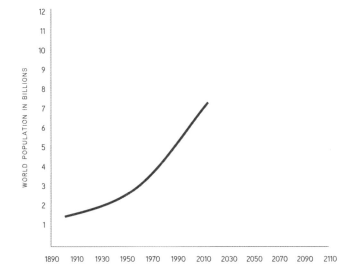

In this figure, we have added the median variant United Nations(UN) projection to 2100.

We see the change from accelerating growth to decelerating growth. We also see that the inflection point of the curve was passed in the last decades of the 20th century and that, as of this writing, we are living in an age of slowing growth, one that may be very different, in terms of environmental limits and human social interactions, from the previous period. In part three, we will explore that difference.

Population projections are based on present knowledge of continually changing trends. As such, they cannot be taken as firm predictions of the future, but they do provide us with a perspective for viewing and understanding the present human situation.

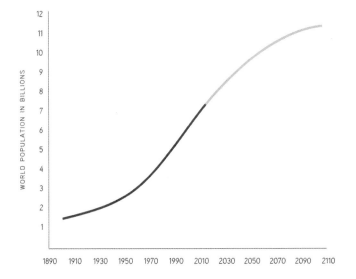

This figure shows the median variant along with the 95%
prediction interval for projected population growth from 2016
to 2100 as estimated by the Population Bureau of the UN.
We can see that the more optimistic projection is a plateau
at a level of under 10 billion, whereas the higher estimate
points to a world population of approximately 13.5 billion.

There is a considerable difference between meeting the needs
of a population of 10 billion and meeting those of 13.5 billion.
Actions we take now will influence the curve to plateau sooner
rather than later. These actions will have a huge impact
on what life will be like in 50 or 100 years. In the coming
pages, we will see that this influence will come through
improvements in social, economic, and environmental
conditions in all regions of the world. Thus, doing what we
can to improve quality of life for all and positively affect the
course of growth is of the utmost importance.

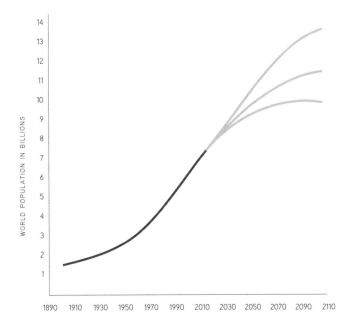

The human population is going through a period of demographic transition. At the top of the figure at right, we see a plot of population size showing the familiar S-shaped curve. The middle figure shows birth and death rates over time. The bottom figure combines the two. The time period described breaks down into four phases. A closer look at these explains the dynamics underlying the sigmoid growth curve.

In the first phase, corresponding to the preindustrial period, birth and death rates are high but approximately equal; population size remains steady. In the second phase, with the modernization of agriculture and industry along with improvements in health care and sanitation, death rates fall. Infant mortality decreases and people live to a more advanced age. As the gap between birth and death rates increases, the overall growth rate increases, and population grows. In the next phase, in response to changing conditions, people have fewer children and birth rates fall. Population growth continues but at a slower and slower rate, and the growth curve bends downward. In the final period, birth rates and death are once again balanced, and the growth curve begins to level off. Population returns to equilibrium.

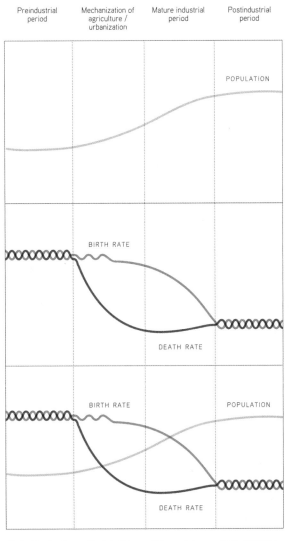

| Preindustrial period | Mechanization of agriculture / urbanization | Mature industrial period | Postindustrial period |

POPULATION

BIRTH RATE

DEATH RATE

BIRTH RATE

POPULATION

DEATH RATE

| Big families, farming, death, and disease commonplace | Food plentiful, people healthy | Baby survival up, women have more work opportunities | People live longer, fewer babies born |

Why does population growth slow?

The dynamics of population growth are complex, but it appears that improvements in health care, lowering of infant and maternal mortality, availability of education—particularly for women—and overall economic development result in people deferring the birth of the first child and having smaller families. The further lowering of birth rates and slowing of growth are thus closely tied to improving conditions throughout the world.

It is a notable phenomenon that reducing problems and increasing well-being accompany slower population growth. We once thought that more people would lead to more problems, but our approach and perspective have shifted. Now we understand that more solutions lead to slower growth.

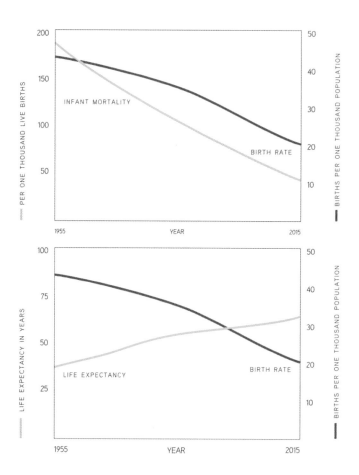

These graphs display crude birth rates in relationship
to variables reflecting human well-being in India.
The two graphs on this page clearly show that birth
rates decline as infant mortality declines and life
expectancy increases.

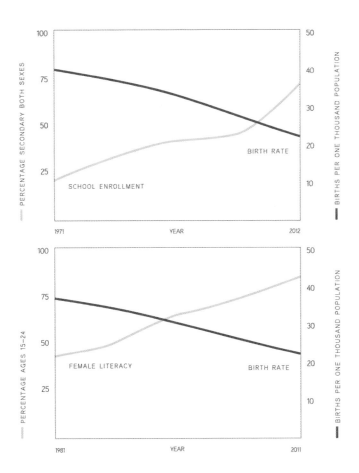

On this page, with respect to education, we see that birth rates decrease as both secondary school enrollment and female literacy rates increase.

Similar patterns are seen in most nations throughout the world. While relationships among such variables are complex, it is now generally accepted that investment in human well-being has the benefit both of improving people's lives and of slowing population growth.

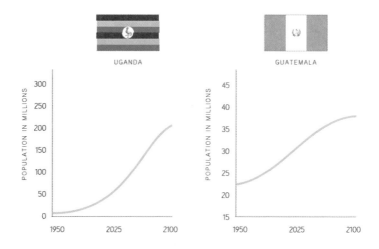

UGANDA

GUATEMALA

POPULATION IN MILLIONS

300
250
200
150
100
50
0

1950   2025   2100

POPULATION IN MILLIONS

45
40
35
30
25
20
15

1950   2025   2100

The transition from accelerating to decelerating growth
occurs at different times, and population size will peak
at different dates in various nations. These figures show
population growth and median variant projections for four
countries, each at a different phase of the demographic
transition. Though each country has a different population
size, we can see from the shape of each curve that the
point of inflection and the plateau of population will occur
at different times.

Some countries, such as Uganda, are just entering the
second phase of the demographic transition; death rates
have fallen, but birth rates are only beginning to decrease.
The inflection point has not been reached. Population size

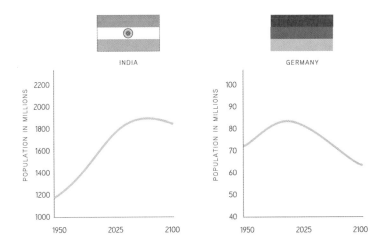

INDIA — GERMANY

will increase dramatically for years to come and will not plateau before the year 2100. Guatemala, however, is well into the second phase; birth rates are decreasing and growth is slowing. India, where birth rates have been falling for some time, is an example of a country well advanced into the third phase. Population is approaching equilibrium and is projected to peak just past the middle of the 21st century.

In some countries in the developed world, such as Germany, population growth has already reached a plateau. As pictured above, this occurred around the year 2000. Germany is in the fourth phase of the transition; birth rates are equal to or less than death rates, and population size is projected to remain steady or decline.

This figure schematically shows population curves for a number of countries based on UN data and projections as of 2010. Each colored band represents the population size of a single country; these are stacked one upon the other, and their sum is the size of world population. In the figure, we can see the relative impact of population size and growth of individual countries within the overall picture of world population. The diversity of the change in terms of timing and magnitude is apparent, as is the similarity of the overall pattern of growth.

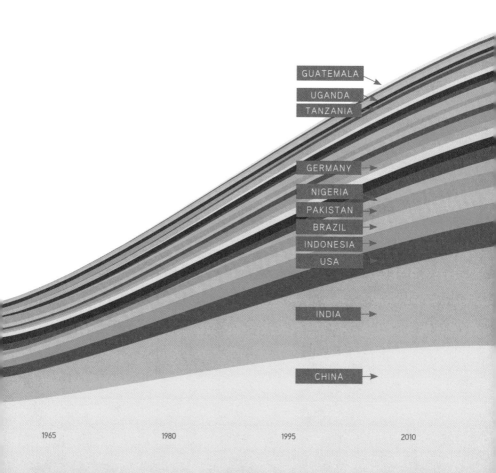

2025   2040   2055   2070

More Developed Countries

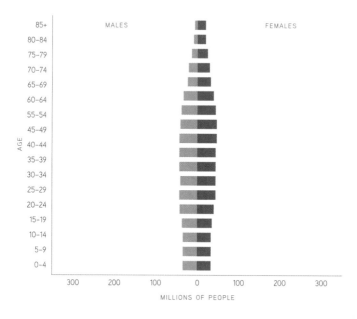

A look at age distribution in the populations of more and less developed regions is also revealing. The narrow-based and straight-sided profile in the more developed regions, shown above, is typical of a stable population. Birth rates are low, and more children are living to an advanced age.

The broad base and narrow top of the profile in the less developed regions, shown on the next page, indicate higher birth rates, a large percentage of the population under the age of 25, and a growing population. As conditions have improved, fewer children are dying in infancy and early childhood. We can expect many of these young people to live longer than in the past. They will then fill out the

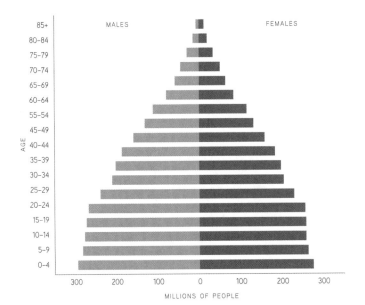

Less Developed Countries

MALES   FEMALES

AGE: 85+, 80–84, 75–79, 70–74, 65–69, 60–64, 55–54, 45–49, 40–44, 35–39, 30–34, 25–29, 20–24, 15–19, 10–14, 5–9, 0–4

MILLIONS OF PEOPLE

300 200 100 0 100 200 300

higher age levels in the figure. Population size in these areas will continue to grow for the next 50 to 100 years. Eventually, the figure for the less developed regions will be straight-sided like the figure for the more developed regions. Population size will be stable, but there will be many more people in what we currently think of as the developing world.

An additional inference from this figure is the huge proportion of young people in the world. This constitutes a whole generation, the largest of any in human history, who are influenced by current events and who will, in turn, influence the course of the future.

If we plot population size in the more developed regions and still developing regions, a striking picture emerges. Population growth in the more developed countries has slowed over the past 50 years and is nearly at a plateau. In the less developed regions, even after birth rates come into line with death rates, population size will continue to increase because of the proportion of young people in these regions. While population growth rates will be falling, population size will continue to increase well into the end of this century. As a result, by 2100 nearly 90% of the human population will be living in what are currently developing countries. As this is occurring, developing countries are taking a larger and larger role in global affairs. Those in the developed world are increasingly taking into account conditions, events, and human well-being in what is now the developing world.

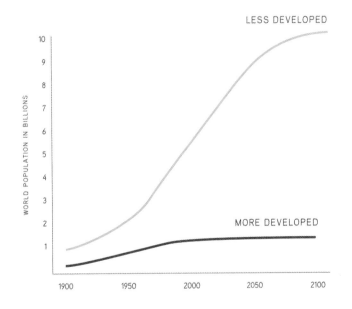

This final figure maps the course of human population growth and population growth rates over a period extending from 8,000 years in the past to 8,000 years in the future. The curve indicates a plateau of world population at approximately 11 to 12 billion by the end of the 21st century and is based on the assumption that fertility in all countries will revert to replacement levels, resulting in a population size that will remain stable or slowly decline.

If these estimates and assumptions are valid, we can see that the present extended period of rapid population growth is unique when seen from a long-range perspective; it has never occurred before and is unlikely to occur again. Our projection of steady-state population size over thousands of years is not certain, but it is useful in depicting the uniqueness of our current situation.

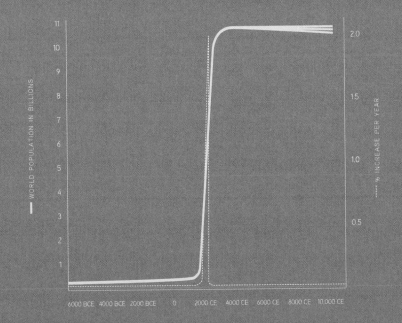

We have seen data and projections that indicate that, while not certain, the human population growth curve may well follow a sigmoid pattern. After hundreds of years of acceleration, worldwide population growth began to slow in the last part of the 20th century. In that period, we passed the region of inflection of the curve, and in the 21st century, we are fully in a period of decelerating population growth.

It should be noted that while a plateau of population size is projected, the level of that plateau is not yet determined. Actions we take now will have a profound influence on the actual trajectory of the population growth curve. Further slowing of growth will be facilitated by improvements in health, education, and economic security as well as access to technology for each person in every nation of the world. Even if such efforts are as successful as we hope, in the coming decades we will still have to meet the enormous challenge, first, of providing for more people than ever before and, second, of adapting to limitations in terms of resources—a set of conditions different from any we have faced in our history.

The conditions of the near future will be different from those of the recent past. The change from increasing to decreasing growth rates represents an epochal change in worldwide population trends. As this occurs, human attitudes, values, and behavior must also change.

PART THREE

A New Epoch

PERIOD I

The long-range course of human population growth can be divided schematically into three periods. First, the very gradual increase of pre- and early agricultural history.

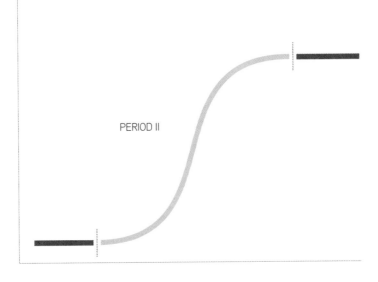

PERIOD II

Second, our current period of rapid change, in which growth follows a sigmoid pattern.

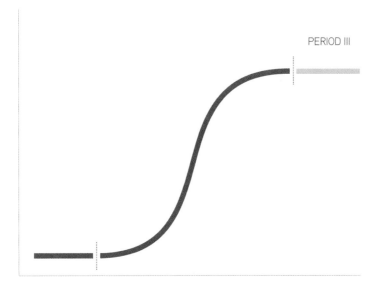

PERIOD III

Third, the more distant future, when
population size will be steady or declining.

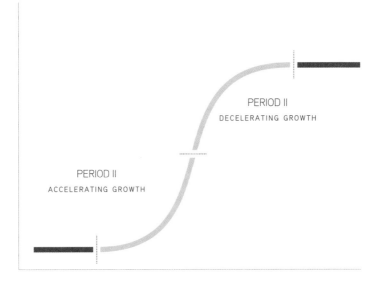

PERIOD II
DECELERATING GROWTH

PERIOD II
ACCELERATING GROWTH

The second phase of the graph can be seen
as two subperiods divided by a point
of inflection: one of accelerating growth
followed by one of decelerating growth.
In this section, we focus on this period of
rapid change, using the schematic of the
sigmoid curve to discuss the nature of human
values, attitudes, and behavior before and
after the point of inflection.

The sigmoid growth curve consists of two sections of different shape: the upturned portion describes a phase of progressive acceleration of growth; the second portion is downturned and describes a phase of progressive deceleration. The difference in shape between the two portions of the curve suggests both quantitative and qualitative differences in human life between the two periods of time. It not only indicates differences in population growth patterns but also suggests differences in the characteristics of prevailing conditions and the nature of human life in the two periods.*

* We are using the sigmoid curve as an image of qualitative as well as quantitative change over time, but in this and the following sections we have not labeled the axes. When the curves are used to reflect quantity, the horizontal axis indicates time and the vertical axis, number. When the curves reflect qualitative differences, the horizontal axis reflects time and the vertical axis indicates change in relative prevalence.

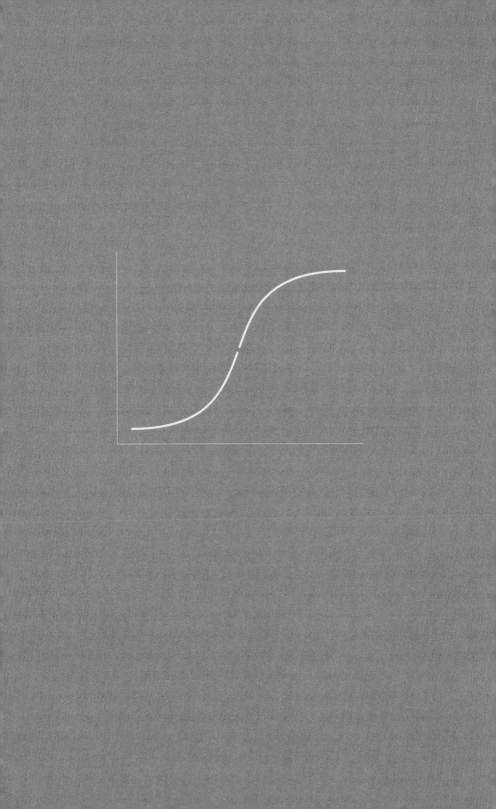

In this figure, for emphasis, the two parts of the curve before and after the point of inflection are separated. We have labeled one A and the other B. The periods of time prior to and following the point of inflection are referred to as Epoch A and Epoch B, respectively.

The orientation of the curves suggests that life might look very different depending on the era in which one is living. To someone born in Epoch A, the future would appear to have few limitations in terms of growth, resources, and available energy. Someone living in Epoch B would, however, have a distinct sense of limitations and of the necessity to adapt to the approaching of a plateau in population growth.

The difference in the shape of the curves implies that there will be fundamental, qualitative differences in the nature of human life between the two epochs.

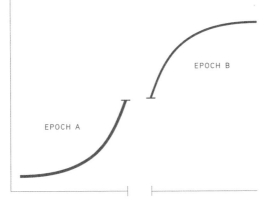

EPOCH A

EPOCH B

Given the difference between the two periods, it is likely that certain attitudes, values, and behavior that were of positive value in Epoch A may be of negative value in Epoch B.

A clear example is population size. In Epoch A, progressive increase in population was seen to be positive; in Epoch B, this increase is now of negative value and, if left unchecked, threatens our very existence. On an individual level, in the past it was economically and socially desirable to have large families. As society changes, it becomes more desirable to limit family size.

In a way that parallels biological evolution, certain human traits and tendencies that were advantageous in Epoch A were selected for. In the differing conditions of Epoch B, there is a shift such that other traits and tendencies will be of greater advantage and become more prevalent.

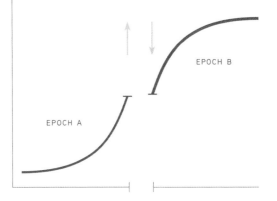

EPOCH A

EPOCH B

Perhaps the clearest example of the shift from A to B is that of resource use. During the period of exploration and expansion by European colonial powers, followed by the period of industrialization in both Europe and the United States, resources seemed limitless. They could be exploited without regard for the effects either of consumption or of the disposal of waste. This would correspond to Epoch A, in which positive value was placed on growth, consumption, and unlimited use of resources.

In the last 50 to 75 years, however, there has been increasing awareness that resources are limited and that unfettered consumption, along with disregard for the effects of waste products, endangers our survival. Our adaptive response has been to place increasing value on awareness of limits, on conservation, and on sustainability.

Thus, the conditions of Epoch A support and are consistent with values of unlimited growth and consumption, while the different conditions of Epoch B will lead to the different values of sustainability and conservation.

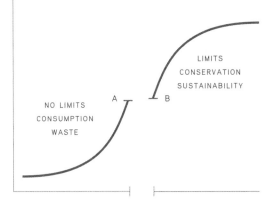

In Epoch A, there was an appropriate emphasis on the quantity of children each family would have. In the changed conditions of Epoch B, in which fewer children will be born to each family, the emphasis can be expected to shift to the quality of care of children, the quality of each child's experience, and the overall quality of human life.

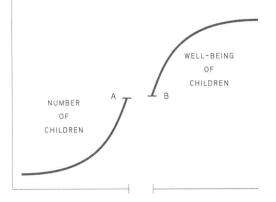

NUMBER
OF
CHILDREN

A

B

WELL-BEING
OF
CHILDREN

During the beginning of Epoch A, because mortality rates were high at all age levels, the control of disease and premature death were of primary concern. Success in this regard has contributed to the recent sharp increase in population size.

As we transition from Epoch A to Epoch B, with advancements in medicine and public health reducing the incidence of disease and premature death, there is an increasing emphasis on prevention, on promoting health, and on enhancing quality of life.

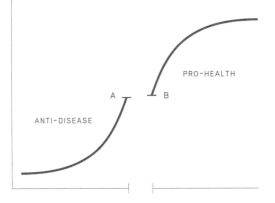

PRO-HEALTH

A

B

ANTI-DISEASE

In Epoch A, expansion was not constrained, and those societies and industries that grew most steadily tended to dominate. This was clearly evidenced in the development of modern industrial nations. As a result, persistent growth and expansion continued to be the dominant influence in modern social and economic life.

The shape of the second part of the curve suggests that in the different reality of Epoch B, an orientation toward equilibrium will be more appropriate than one toward persistent expansion. Population, material production, and consumption are expected to reach a plateau. This plateau, however, will not necessarily be a period of stagnation. It will be a time of dynamic equilibrium in the material realm and continued development in the human realm. While there will be a plateau in population size, there will be continued improvement in the quality of human life.

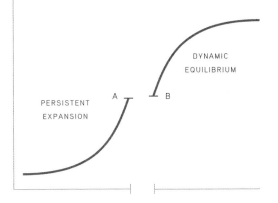

Similarly, qualities of competition, independence, and use of power were successful and tended to dominate in Epoch A. In the different conditions of Epoch B, strategies involving collaboration, interdependence, and consensus will likely be of more value in resolving conflicts and providing for basic human needs.

We have recently seen evidence of this shift toward collaboration and consensus in the accord signed by 191 nations with respect to climate change. This is a momentous agreement that is emblematic of the shift from Epoch A to Epoch B.

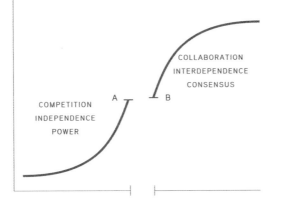

COLLABORATION
INTERDEPENDENCE
CONSENSUS

COMPETITION
INDEPENDENCE
POWER

A

B

With the development of digital technology and the Internet, which came into being at the beginning of Epoch B, another value shift is occurring: human beings changing from a state of being loosely connected through the last decades of Epoch A to being more tightly and instantaneously interconnected in Epoch B.

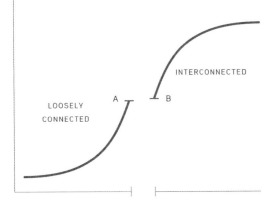

In Epoch A, competition and the demands of persistent, accelerating growth were inherently associated with either/ or attitudes and philosophies and the prevalence of win-lose strategies in the resolution of conflict. People or nations saw the world as a place in which any benefit to the other is a loss or detriment to the self. In Epoch B, however, the tendency toward balance, collaboration, and interdependence will be based upon and evoke a philosophy of both/and and the development of win-win strategies.

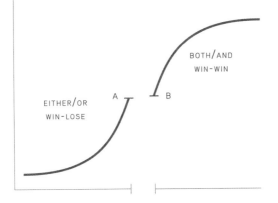

In Epoch A, inherent in the process of accelerating growth and change is a tendency toward extremes. Thus, excesses in growth and development and in the use of natural resources often occurred. In Epoch B, however, a tendency toward balance can be expected as part of the process of slowing growth, and balance will become evident both in relationships among human beings and in relationships between human beings and nature.

For example, in the conditions of Epoch A, unrestricted use of natural resources was practiced. However, as conditions change, excessive use of resources is revealed to be unwise.

The trend toward excess has resulted in an imbalance in distribution of wealth and resources. As the rate of growth diminishes, reduction of extremes and increase in balance will occur. The overall benefit to society of a more balanced distribution will be of advantage to both individuals and nations, rich and poor.

The tendency toward extremes has, throughout the world, also led to our species being out of balance with nature. As we adapt to Epoch B, that balance will need to be restored.

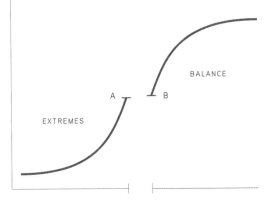

In Epoch A, short-range benefits and costs were of greater concern than long-range. Orientation was toward the present, and the effects of actions were usually seen in isolation—as parts unrelated to other human groups or the whole of the ecosystem. In Epoch B, limitations of resources and high population density will result in an increased tendency to approach problems with a focus on the future, on long-range effects, and the understanding that we, as individuals, as groups, and as a species, are part of a larger whole.

One example is seen in the effects of waste disposal. In early phases of industrial development, the adverse effects of waste either were not perceived or could be ignored. In Epoch B, however, in part because of increased density of population and increased production of waste, the consideration of such adverse effects will be essential. Avoidance or amelioration of pollution will be in the interests both of neighboring populations—either human or other species—and of those who may be producing the waste.

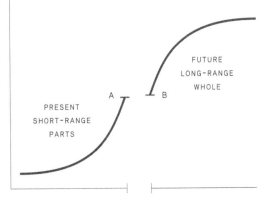

PRESENT
SHORT-RANGE
PARTS

A

B

FUTURE
LONG-RANGE
WHOLE

The final figure of this section is intended to underscore the nature and significance of the change that is occurring in our time and to suggest the emergence of a new reality in the history of humankind.

A major shift in our perception of reality occurred in the past, when a prior belief that the earth was flat was altered by evidence that the earth was, in fact, round. Another shift occurred when the belief that the sun revolves around the earth was changed by the discovery that the earth revolves around the sun. These changes in our relationship to the earth and sun were changes of perception; the realities were not altered, only our perception of them.

Now, however, another major change is in the making. This is a change from seeing the world as limitless in terms of growth to seeing it as limited. It is also a change from seeing ourselves in opposition to each other to seeing ourselves in collaboration with one another. It is due not only to a change in perception but also to a change in external reality as expressed by the shape of the curve. The change from A to B can be seen in our relationships to nature, our relationships to one another, and in our relationships to ourselves.

RELATIONSHIP OF:

HUMAN TO EARTH

HUMAN TO SUN

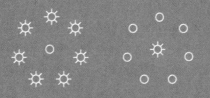

HUMAN TO HUMAN

Humans can see future consequences of
their actions and adapt their behaviors
accordingly. Making these adaptations will
require much striving and work in the decades
to come, but it is essential that we start now.
Just as shifts in population growth patterns
occur over time, qualitative shifts in attitudes,
values, and behavior can also take place
over time—likely several human generations.
This brings us to the realization that,
individually and collectively, we share the
responsibility for the future course of events
on this planet, whether these events are
positive or negative. This outcome depends
on the way in which we respond to the new
reality of population change and available
resources. The images presented suggest
the possibility, as well as the necessity,
of responding to these changes in a positive
and humane way.

We can see, however, that in this crucial
period of change, there will be conflict
between differing systems of values.
A difficult period of transition lies ahead.

The figures in this section suggest that accompanying the change from accelerating growth to decelerating growth, a major shift must occur in human values, attitudes, behaviors, and relationships.

PART FOUR

Paradox and Conflict

Human beings possess the capacity for a wide range of attitudes and behaviors. The idea underlying the discussion in part 3 is that attitudes and behaviors that are appropriate in the reality of Epoch A may be less advantageous and less appropriate in the reality of Epoch B. Another set of values that is less advantageous in Epoch A may be much more so in Epoch B.

The context or circumstances that prevail determine which attitudes and behaviors are appropriate at different times. Thus, the emergence of Epoch B values outlined in part 3 is seen as a necessary response to the different reality of that era.

In the region of inflection, growth rates are highest; acceleration is changing to deceleration; and values are shifting most rapidly. The period can be expected to be a time of increased conflict. In the following section, we will look more closely at this period.

The change described in part 3 will be the result of a kind of natural selection for human attitudes, values, and behaviors. We, as human beings, possess the capacity for both Epoch A and Epoch B behaviors. In the conditions of Epoch A, Epoch A behaviors will be the most advantageous and will predominate. In the different conditions of Epoch B, Epoch B behaviors will confer the most advantage to those who exhibit them. Those values, attitudes, and behaviors will be adopted and will predominate or become integrated with those of Epoch A.

Looked at in this way, we can see that the changes in values described in the preceding section are the result of learning, adaptation, and a process of social evolution.

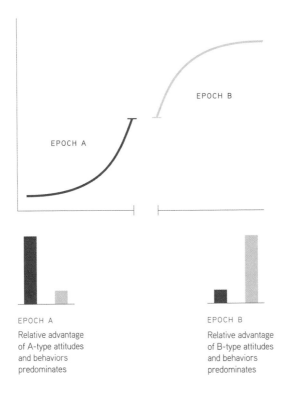

EPOCH B

EPOCH A

EPOCH A

Relative advantage
of A-type attitudes
and behaviors
predominates

EPOCH B

Relative advantage
of B-type attitudes
and behaviors
predominates

This change in values is based not necessarily on "right" or "wrong" but on what is best for the individual. As seen in this figure, the attitudes, values, and behaviors that serve self-interest in Epoch A are very different from those that do so in Epoch B.

Paradoxically, this means that in the reality of Epoch B, behaving in a more generous, community-oriented manner will better serve the interest of both individuals and groups than behaving in an exclusively self-oriented, competitive way. Therefore, collaboration, awareness of others, and balance will not be regarded as personal sacrifice; instead, they will be personally and collectively beneficial.

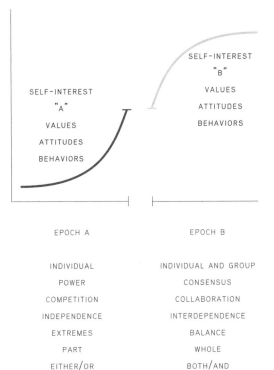

SELF-INTEREST
"A"
VALUES
ATTITUDES
BEHAVIORS

SELF-INTEREST
"B"
VALUES
ATTITUDES
BEHAVIORS

EPOCH A

EPOCH B

INDIVIDUAL

POWER

COMPETITION

INDEPENDENCE

EXTREMES

PART

EITHER/OR

INDIVIDUAL AND GROUP

CONSENSUS

COLLABORATION

INTERDEPENDENCE

BALANCE

WHOLE

BOTH/AND

In the context of Epoch A, the generous or humane attitudes appropriate in Epoch B were not perceived as pragmatic. However, in the different reality of Epoch B, such attitudes will be both pragmatic and humane.

Thus, the shift will not come about simply because Epoch B values are morally or spiritually better than those of Epoch A. They will change because the values of Epoch B, in the context of that epoch, will be more advantageous.

For example, improvement in the quality of life in developing regions and the self-sufficiency of those nations will benefit the people both in those areas and in the more developed world. Improvements in health care, education, and economic viability in the less developed areas will help in ameliorating population pressures, which would benefit the world as a whole. In addition, a balanced relationship of wealth and exchange would lead to more stability as well as an increase in personal satisfaction and well-being. In Epoch A, such changes would have been perceived as conferring no advantage to the interests of the more developed areas; now they are being seen as advantageous to all regions.

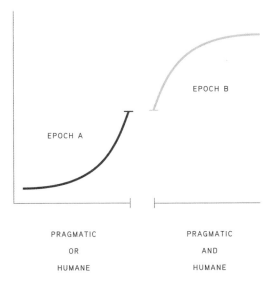

EPOCH B

EPOCH A

PRAGMATIC
OR
HUMANE

PRAGMATIC
AND
HUMANE

Returning to the population growth curve, we get an interesting view of this process of change. Each generation comes into being at a different point on the curve.

Here we draw in the life span of the Depression generation, the Baby Boom generation, and the Millennial generation—the last being those born during the last two decades of the 20th century.

We see that people born in the years preceding the Great Depression lived out their lives almost entirely in Epoch A, in the part of the curve where population growth was accelerating. Those in the Baby Boom generation were born just before the point of inflection; however, the inflection of the curve occurred during their lifetimes. Thus, they were born in the reality of Epoch A but have lived the later part of their lives in Epoch B— the part of the curve where growth is slowing.

Those in the Millennial generation were born after the point of inflection of the growth curve, fully in Epoch B. From the time of their birth, the reality they have experienced has been one of awareness of limits, the need to conserve, and the sense of the planet as an integrated whole. Thus, their attitudes, values, and behaviors have been shaped by and are adapted to a reality very different from that experienced by any generation before them.

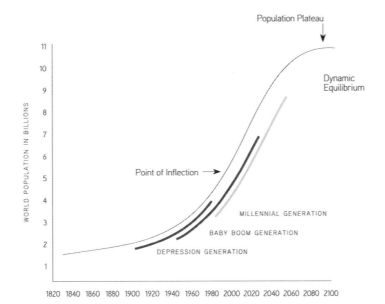

Noting this, we can see more clearly how this transition is coming about. For those born early in the 20th century (and before), their perceived reality was of limitless, unencumbered growth; they perceived and adapted to the reality of Epoch A. Those born after World War II came into being in the conditions of Epoch A—seeing uninterrupted growth and limitless potential. However, partway through their lifetimes, in the 1960s and 1970s, conditions changed. Limits became apparent; the effects of pollution could not be ignored; energy crises and shortages of oil occurred; eventually, in the 1990s, the reality of climate change came into focus. Because of this, increasing value began to be placed on conservation, on control of toxic waste products, and on balance with the natural environment. But this generation was caught in the middle, showing qualities of both Epoch A and Epoch B.

From the time the Millennials were small children, the reality of limits, the perception of the world as an integrated whole, and the need to cooperate in the use of limited resources have been part of their experience.

Thus, over a period of three generations, reality will have changed and with it the predominating values.

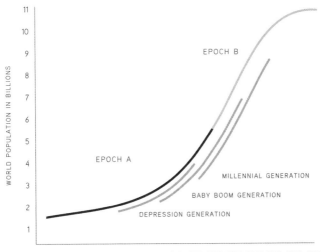

The region of inflection is a time of transition from the predominance of Epoch A values to that of Epoch B values. In this figure, the relative positions of the two lines indicate that Epoch B values exist even in the period before inflection but are less dominant than Epoch A values. In Epoch B, the relative dominance is reversed. As indicated, the tendency of one set of values to persist as the other begins to emerge will give rise to conflict and uncertainty.

This figure offers an explanation for the tension we feel at this time. It suggests that the conflicts are an inherent part of this developmental and evolutionary process. They are not necessarily a sign of the impending end of the human species but instead reflect the process of inversion in values that is now occurring.

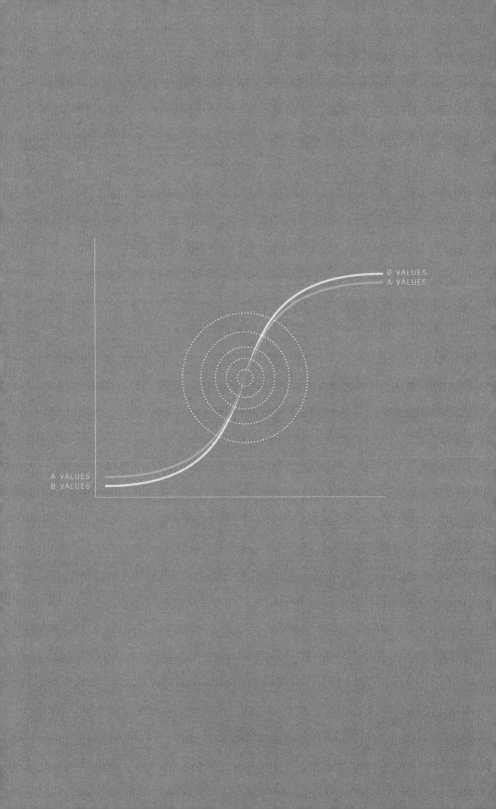

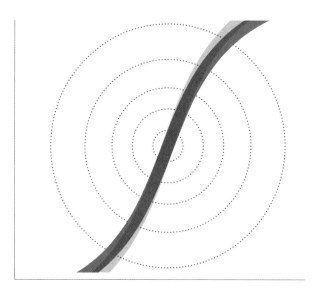

When viewed from a short-term perspective, as represented
in the figure above, the tension and conflict inherent in
this transition may seem chaotic and symptomatic of
a disintegrating, collapsing world.

However, when viewed from a longer-range perspective, as shown in the sigmoid curves in this figure, these conflicts and uncertainties can be seen as part of an orderly if somewhat difficult process of nature. Looked at in this way, the disturbances of the present time may be seen not as a symptom of a disease that must be treated or eradicated but as a result of the obsolescence of formerly successful patterns of life and the uncertain beginnings of new patterns appropriate to the emerging conditions.

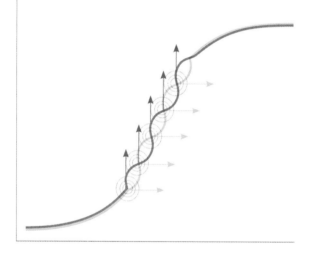

Over an extended period of time and with the dominance of
both/and attitudes, the transformation from Epoch A to Epoch B
can be seen as a series of reconciliations of opposing tendencies
and as part of a continuing process, as suggested in the figure
above. Conflicts are apparent, but their synthesis is facilitated
by both/and strategies. Bringing about this kind of resolution
is one of the great challenges of the present era. Success in this
respect will result in the evolution of new ways of life that will
be beneficial both to individuals and groups and to the furthering
of human evolution.

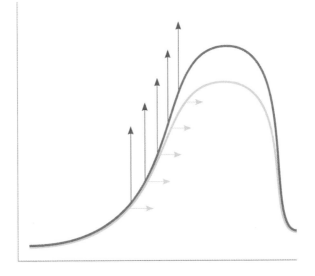

The continued dominance of either/or attitudes and Epoch A approaches, however, will lead to an escalation rather than a reduction of conflict as depicted in the figure above. In time, this would lead to widespread unrest, depletion of resources, and possibly a total collapse of the human population.

The effective use of the both/and approach is symbolized here. As conditions approach population equilibrium, there will be a progressive rapprochement and integration of Epoch A and Epoch B values.

For example, a complete disregard of the technological and social developments of Epoch A in an effort to immediately halt growth will be inappropriate and unrealizable. However, attempts to resolve tensions by completely suppressing the tendencies of Epoch B will be equally disadvantageous. With a both/and approach, the developments that have been part of Epoch A can be combined with Epoch B values in order to develop solutions that are appropriate to this era and carry us to the plateau of the population curve.

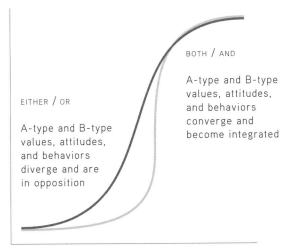

EITHER / OR

A-type and B-type values, attitudes, and behaviors diverge and are in opposition

BOTH / AND

A-type and B-type values, attitudes, and behaviors converge and become integrated

We have suggested that Epoch B values, attitudes, and behaviors are emerging not only because they are humane but also because they are advantageous to individuals and to society. During this transition, it can be expected that conflict, at all levels of human life, will increase. In the long term, such conflict will be most effectively and constructively resolved with both/and rather than either/or strategies and through the integration of the values of Epoch A and Epoch B.

The present period is especially sensitive. In resisting change, we may cling to values that are obsolete and exceed the tolerance of nature. Resisting change may ameliorate some problems in the short term but will not provide the basic shift in values needed in this epochal transition. Strategies that resist change greatly increase the risk of disaster through famine or armed conflict.

Transition to a new epoch will be achieved through the creativity, initiative, and shared responsibility of individuals throughout the world and through the successful integration of conflicting tendencies.

Though the present period of crisis confronts us with the danger of self-extinction, it also presents us with an opportunity for development that could aid long-term survival of the human species and enhance the quality of individual life.

PART FIVE

Resolution and Integration

In the preceding sections, we have presented shifts in human society and human values that will need to take place if we are to transition successfully to Epoch B. The challenges of this transition are substantial, and meeting them will involve change at every level of human existence, from the individual to the global.

In this concluding section, we will indicate that the resolution of conflicting tendencies in the human realm has begun and will likely continue. The convergence and integration of the values of Epoch A and the values of Epoch B will result in an as-yet-to-be-created synthesis of the two.

We are at a unique juncture in human history and evolution. We are faced with the opportunity and the responsibility to create new relationships, systems, and institutions. This will come about through the adaptive convergence and integration of currently divergent and conflicting tendencies.

This figure, which was first presented in part 2, reveals an image of the present era when seen in the long-range context of population growth.

We see that the human species is experiencing far more rapid and sustained change in terms of population growth and sociocultural development than has occurred before or is likely to occur in the future. Intuitively, we sense from this image that, from a biological, psychological, and social standpoint, human beings may be better adapted to conditions associated with less rapid change (such as those that existed in the more distant past and that are anticipated in the coming centuries) than they are to those we currently experience.

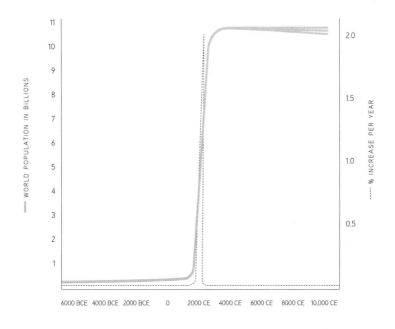

In this critical period, we are responsible for guiding our society and our species to a new equilibrium.

As mentioned earlier, the population growth curve may be viewed as consisting of three sections: one, a period of very slow growth; two, the current period of rapid growth, which follows a sigmoid shape; and three, a future period of relatively slow change but with a much larger population.

It is interesting to note that, because of the large population size and the development of technology, life in Period III will be similar to our present experience. However, in other ways, such as low growth rates, awareness of environmental limits, and tendency toward equilibrium, some of the qualities of life in the far future will resemble those that existed in the distant past—that is, Period I— more than those in the present and in the recent past—Period II.

This figure gives rise to the idea that our adaptations to the future may be in part the result of the reemergence of qualities that were a significant part of life early in our evolutionary history and their integration with qualities that have recently prevailed.

PERIOD III

PERIOD II

PERIOD I

VALUES, ATTITUDES, BEHAVIORS		
Equilibrium	Growth and Expansion	Equilibrium

TECHNOLOGY		
Tools	Machines	Information Technology

ENVIRONMENTAL CARRYING CAPACITY		
Stable	Increasing Then Stable	Stable-Declining

GROWTH RATES		
Low	High	Low

VALUES		
A+B	A > B or B > A	A+B

Epoch A and Epoch B values, in balance in the distant past, diverged in the upturned part of the curve, with Epoch A values predominating. In the downturned part of the curve, Epoch B values return and are gradually integrated with those of Epoch A in the very different conditions of the future. The figure here schematically represents this.

This reconciliation of conflicting tendencies will be the hallmark of the coming decades. New relationships, new communities, and new modes of interacting will emerge and be integrated with developments in science, technology, economics, the arts, and international relations.

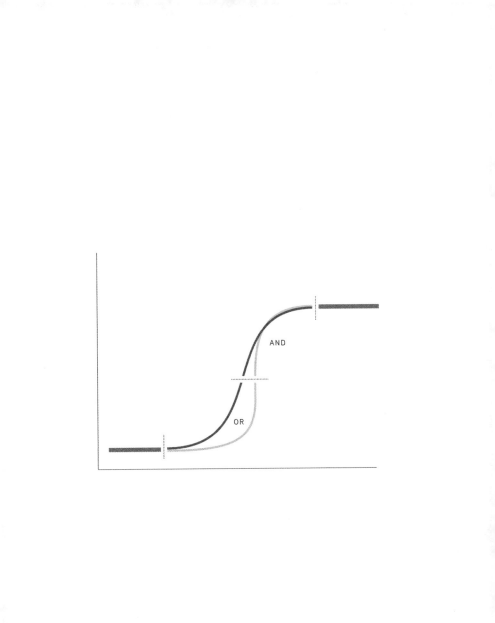

It is a remarkable time that, if we are wise and make the necessary changes, can be one of excitement, creativity, and extraordinary innovation in both the human and the technological realms.

Epoch A was a period of technological development like no other seen in human history.

This development will continue in the foreseeable future, but the integration of these advances with emerging human values will be necessary for successful adaptation in the decades to come.

In Epoch A, there was often conflict between what was of material value and what was of human or social value. Industries tended to exploit both natural and human resources. Profit, wealth, and unremitting growth often proceeded with little regard for the well-being of individual human beings or the conservation of natural resources. Our task in the future will be to reconcile human value with material value in a way that supports the overall quality of life for all human beings.

As part of this process, there is a need for an approach to economics that places value on both material and human, as well as a need for the emergence of economic relationships that minimize the exploitation of one group or segment of society by another.

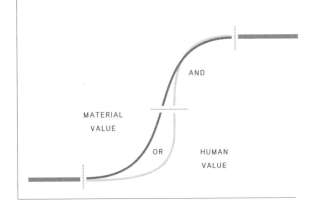

MATERIAL
VALUE

AND

OR

HUMAN
VALUE

In the period to come, the needs and welfare of the developing and developed countries will be more closely intertwined than ever before. Healthy and stable development in all countries will have a positive effect in terms of population pressures, political stability, and sustainability. The achievement of these goals will be of sufficient benefit to the developed nations that it will be in their self-interest to provide support to the developing regions.

It is important to note that significant gains have been made in these areas, exemplifying the change to Epoch B. In recent decades, economic development has proceeded and birth rates have fallen. Furthermore, levels of poverty have decreased and levels of overall health have increased throughout the world. There is an awareness of the importance of the relationship between the developed and developing worlds at the international level, in business, and in the focus of philanthropic foundations.

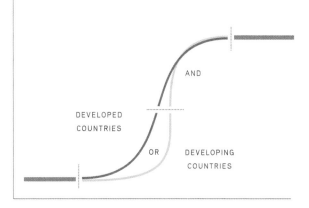

DEVELOPED
COUNTRIES

AND

OR

DEVELOPING
COUNTRIES

For much of its evolutionary past, humanity lived in hunting and gathering or small-scale agricultural societies. These cultural milieus—what we call here a traditional way of life in terms of child rearing, care of the elderly, conflict resolution, family relationships, and sustainable adaptation to the ecosystem—fit conditions of relative equilibrium. With the development of cities and nations as well as modern agriculture, industry, and technology, many of these more traditional ways of life have been overshadowed by modern tendencies. As we move into Epoch B and the anticipated period of equilibrium, some of the elements of traditional cultures will be essential, in combination with those of the present, in the creation of altogether new societies for the future.

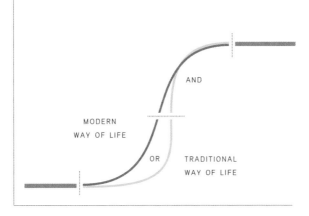

MODERN
WAY OF LIFE

AND

OR

TRADITIONAL
WAY OF LIFE

In the future, with both high population density and increased emphasis on the quality of individual life, forms of social and political organization must develop that are responsive to the needs of both individuals and large groups. Thus, an opportunity is now emerging for integrating local, small-scale organizations—which provide for the political, cultural, and material needs of the individual— with large-scale, centralized organizations concerned with communication among groups and with global coordination of efforts to meet the needs of societies throughout the world.

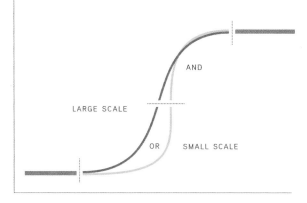

As we have approached limits, a conflict has emerged between the human tendency to consume and the need to conserve. The integration of these two has resulted in the concept of sustainability.

Sustainability has emerged as a concept and a goal in such varied areas as international development, agriculture, industry, and architecture. Within the world of architecture and design, the goal of sustainable buildings and communities has become a highly valued consideration in the planning and creation of new environments. Businesses and corporations are incorporating sustainability into their short- and long-range plans. Developed and developing nations alike are including sustainability in planning.

Arriving at a sustainable future demands that our social, economic, political, and international relationships support the further development of sustainable practices on a large scale. Our overarching task is how to bring this about in a way that is not only economically feasible but also economically beneficial.

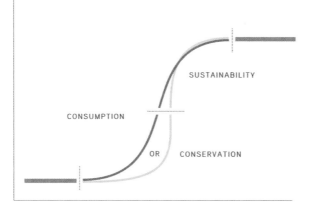

CONSUMPTION

SUSTAINABILITY

OR   CONSERVATION

In the conditions of Epoch A, there have been a divergence of and a specialization in differing disciplines or modes of thought when it comes to human endeavor. There is a distinction and even conflict between science and art, intuition and reason, emotion and cognition. These distinctions extend from day-to-day life to areas of academic research and analysis to the making of governmental and social policy. In the coming century and beyond, we will see a reconciliation of these divisions and the development of integrative, interdisciplinary approaches to human problems, human thought, and human creativity.

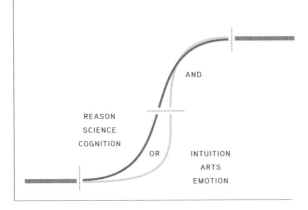

REASON
SCIENCE
COGNITION

AND

OR

INTUITION
ARTS
EMOTION

The adaptive changes described here will necessarily come at every level of human existence, from the molecular functioning of cells and nervous systems to the development of individual thought and emotions, the structure of families, the organization of community and work environments, the social, political, and economic organization of larger societies, as well as interactions among cultures and nations, and the relationship of our species to the planet as a whole.

Approaches to human problems now require attention to and understanding of all levels, and solutions will come through the increasing integration of previously distinct disciplines.

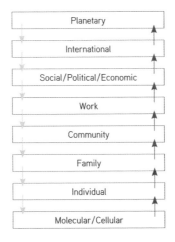

We will conclude with brief descriptions of how this change is occurring and what we might expect in the decades to come.

It is now well known and understood that changes in human thought, feeling, and behavior are accompanied and mediated by changes in cellular and molecular interactions in the human brain and the entire human organism. The changes and adaptations from Epoch A to Epoch B will take place not just at the social level; they will involve, affect, and be affected by the biology—at the level of molecules, cells, and tissues—of all of us and our descendants.

It is also well understood that while each individual's makeup is very much influenced by his or her genetic heritage, the actual phenotype—the body, brain, mind, and person—is shaped by the environment in which he or she grows and lives. Just as the language any human being learns and speaks is determined by the environment in which he or she develops, the social language of values, attitudes, and behavior is shaped by the combination of genetic heritage and the interpersonal milieu in which the individual grows.

Children learn attitudes, values, and behavior based on their cultural and family experiences. In the transition we are depicting, Epoch B values of cooperation, interdependence, collaboration, and win-win conflict resolution will have to be incorporated into family life, early education, and school programs. The manner in which this is done will vary from culture to culture, but this basic shift must take place if we are to survive.

Changes in community relationships will also occur. Incorporating values of collaboration, cooperation, and interdependence will likely result in richer, more complex social networks in terms of mutual support, sharing of child care, setting community goals, and responding to adversity. Doing this in a world of nearly 10 billion people will require as-yet-unknown developments in housing, in urban planning, and in local governments.

Architecture and design will play a significant role in the transition from Epoch A to Epoch B. The design and creation of environments that embody and foster the integration of B values into our lives is a challenge that many in the field understand and will address. Over a generation of architects, the goal of creating sustainable buildings and communities has become a highly valued consideration in the planning and building of new environments.

Economic and business models are also evolving. Enterprises based on the reusing and sharing of existing resources, models that did not exist 10 to 20 years ago, now are enjoying success— an example of the creative incorporation of an Epoch B value (reusing resources) with a quality of Epoch A (a profit-making business model). Other models include using some portion of profits to support segments of the world population that are in need; this represents an increasing expectation that the carrying-out of good business includes acts for the social good.

Most important will be the development of economic models and systems that are adapted to a world with limits in terms of population growth and resources. These models will be different from those of the recent past, which were based on ongoing growth and expansion. It is possible that such models and practices will have the effect of reducing inequalities, resulting in improved health and well-being across all levels of society.

The challenges of the transition are of such magnitude that they can be met only if approached in a cooperative, interdisciplinary manner by varied groups of people.

For example, the problems of reducing poverty, enhancing health and education, and promoting sustainable economic development have to be addressed by collaborators from the fields of economics, international relations, agriculture, education, medicine, and technology. Those who can facilitate cooperation across huge cultural, political, and geographic gaps will also play a vital role. It is important to note that such problem solving cannot be planned and implemented solely by those in the Western, developed world. Collaboration, cooperation, and mutual understanding of local traditions, beliefs, values, and cultural practices will be part of the effort. This human-to-human problem, while difficult to resolve, must be addressed.

In medical care and public health, there is increasing interest in and emphasis on prevention and health promotion as opposed to fighting and attacking disease. It is becoming clear that prevention and health maintenance, in the long term, increase quality of life and are economically wise.

A major change is in communication. The Internet arrived just as we entered Epoch B. Its existence means that small ideas can spread quickly. This is a remarkable difference from our world of 35 years ago. A particular idea at a particular moment in time can affect the thoughts, values and beliefs of millions of people.

At the same time, the Internet provides ways for individuals and small groups to share ideas and creativity. People from vastly different cultures, thousands of miles apart, can interact, develop an understanding of one another, and collaborate in creative ways.

We have never before had such capabilities. Whether this remarkably powerful resource can be used in the service of the transition to Epoch B and beyond remains to be seen. Like all technology, it can be used in the interests of terror, destruction, and zero-sum thinking, or it can be used in the interests of cooperation, creativity, and interdependence. Incorporating it as a resource in the transition to Epoch B is an evolutionary task that we face.

Ultimately, as a species, we have to adapt and live in balance with the constraints and opportunities of nature and of the planet. The integration of values of balance, moderation, conservation, and sustainability, as well as seeing the long-term consequences of actions and understanding the wholeness of the ecosystem in which we exist, is the framework in which we will develop our family, community, cultural, social, economic, political, and international institutions and practices of the future.

In this final figure, we see that our future is something to be designed and created. The challenge of the current and coming generations is to develop different ways of relating socially, economically, and politically from how we have in the recent past—ways that will make use of recent biological and technological advances and integrate them with the values necessary for the near and more distant future.

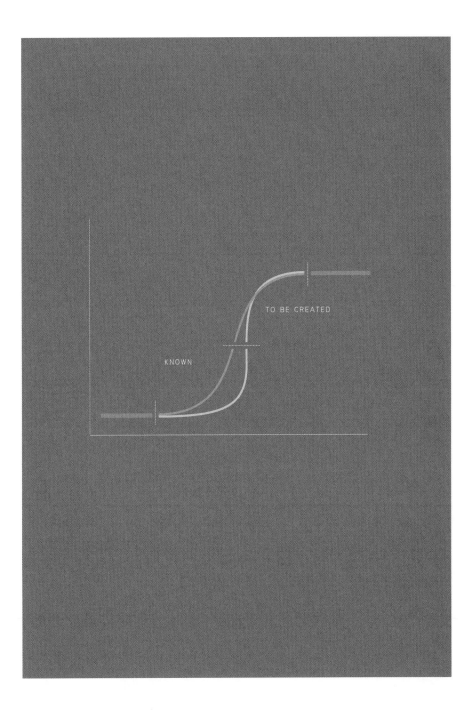

While this task is daunting,
it is also filled with hope.
We have the opportunity,
in fact the obligation,
to react creatively to
our changing conditions
and, if successful, to
survive and live fully for
generations to come.

The epochal change now taking place affects every aspect of human life—individual and institutional, emotional and cognitive, personal and technological. It calls for the resolution of imbalances and conflicts that have arisen in the course of preceding centuries and for the integration of divergent tendencies in human life.

This integration will occur in ways that will differ according to local history, culture, and ecological conditions, but it must occur. Through the human capacity for creativity, through variation among human beings and human societies, and through a process of social evolution, the scientific and technological developments of recent centuries can be reconciled with the values essential to human survival on the planet.

Whether we will successfully make this transition remains to be seen, but the ideas and images presented in these pages suggest our capacity to do so.

The current generation of young people, the largest in history, has the opportunity to shape the future and make it one with higher levels of human health, well-being, and fulfillment than at any time in our existence.

A Favorable Outcome

The epochal transformation described in these pages is occurring through myriad changes at every level of human life. As conditions shift, each generation is born into circumstances that are different from those experienced by the preceding generation. Through different experiences, each has different viewpoints and develops different values. Over a period of time, as conditions change, values, relationships, and systems evolve. As we adapt to the future, the task before us is to allow new values to emerge and to modify our social systems and institutions appropriately.

As population growth slows, and as we approach a plateau in world population size, our greatest challenges lie in innovation not only in the technological realm but, significantly and necessarily, in the human and social realms. Human and social challenges—improving quality of life, feeding billions of people, avoiding disastrous depletion of resources, and creating societies that meet the material and cultural needs of the individual—that now seem insurmountable may in time be no more insoluble than previously "impossible" challenges: development of heavier-than-air flight, modern agriculture, electronics, space travel, and the mapping of the human genome.

At this point, innovation in the area of human development and social relationships is as important as the advent of agriculture 10,000 years ago or the understanding of microbes, molecules, machines, and digital technology in the past century. Just as some of the brightest minds of the past have turned their attention toward the advancement of science and technology and the prevention and cure of disease, many of the brightest minds of the present and coming generations will turn their attention toward the phenomena of the human mind, human relationships, and the improvement of the quality of human life.

It is now clear that this process has begun. Some of the most creative and intelligent women and men of today are confronting the human issues of conflict resolution, poverty, and paths to sustainable economic development. Others are addressing the technology necessary for our survival into the future. There is an emergence of evolutionarily wise and adaptive strategies that are appropriate to the changing conditions.

Because this change is taking place at every level of human life, each of us has a role. There are those who will incorporate new values into their families or communities or places of work. There are others who will shape policy at local, national, international, and global levels. There are those who will make technological and social innovations in science, in communications, in business, in health promotion, and in economic, sociocultural, and international relationships. In doing so, each of us is responding to the necessity that we evolve, and each is contributing to the long-term survival of the human species.

Even though the charts and diagrams suggest that, in the long term, the course of the epochal change is predetermined, it is not. It is subject to our influence. Our survival is not ensured. An urgency exists today that did not when this book was originally written 35 years ago. The reality of an increasing population is upon us; population pressures threaten our social, agricultural, and economic systems. Armed and unarmed conflict is frequent and threatens stability both locally and globally. Human-influenced climate change is only a handful of years away from a point of no return. Taking care of the health and well-being of human beings in every part of the planet is now not just a matter of humanitarian spirit; it is a matter of necessity.

What we do now and how we adapt will have a significant effect on the future. Our actions now will affect the shape of the population curve; our actions now will influence the state of health or disease world-wide; our attitudes now will determine how conflict is resolved in a highly populated world; our values now will affect how quickly we adapt strategies of sustainability; our behavior now will affect how much carbon we put into or take out of the atmosphere. Our actions now and in the coming decades will determine whether we make the necessary changes in human-to-human relationships to ensure our continued existence on the planet.

The future is simultaneously brighter and darker than it was when the first edition of this book was published. Gains have been made. At the same time, as in so many times in history, we are faced with dark forces of conflict, war, mass killing, and terrorism. Our strongest weapon against these is the promotion of health, hope, and fulfillment for all human beings. As difficult and as daunting a task as that may seem, it is one that, if undertaken successfully, will result in a better world for all.

We are at a point in the course of human social evolution when the demands of survival converge with the higher ideals of humankind and the well-being and flourishing of human society. It is up to us to see that we navigate this transition, adapting to and emerging in a new reality.

Frontispiece: From Jonas Salk and Jonathan Salk, *World Population and Human Values: A New Reality* (New York: Harper & Row, 1981), fig. 29, p. 67.

World population size: Population size in 2100 from United Nations (2015). Probabilistic Population Projection based on *World Population Prospects: The 2015 Revision*, Population Division, DESA. http://esa.un.org/unpd/wpp/. Stabilization at plateau and future projection assumes fertility rates in all nations return to and remain at replacement levels.

World population growth rates: Peak growth rate of 2.06 from United Nations, Department of Economic and Social Affairs, Population Division (2015). *World Population Prospects: The 2015 Revision*, DVD Edition. Projections 2015–2100, medium fertility variant.

Part One

Pages 42–43: Adapted from John McHale, *The Future of the Future* (New York: Braziller, 1969), fig. 1, p. 58. Data 1970–2015 from United Nations, Department of Economic and Social Affairs, Population Division (2015). *World Population Prospects: The 2015 Revision,* custom data acquired via website. http://esa.un.org/wpp/DataQuery

Page 55: Raymond Pearl, *The Biology of Population Growth* (New York: Knopf, 1925), fig. 12, p. 35. (Copyright 1925 by Alfred A. Knopf, Inc. and renewed 1953 by Maude de Witt Pearl.)

Page 57: Ibid., fig. 4, p. 9.

Pages 58–59: Adapted from Oscar W. Richards, "Potentially Unlimited Multiplication of Yeast with Constant Environment and Limiting of Growth by Changing Environment," *Journal of General Physiology* 1 (1928): 525–38, fig. 4, p. 534.

Page 61: J. Davidson, "On the Growth of the Sheep Population in Tasmania," *Transactions of the Royal Society of South Australia* 62, no. 2 (December 23, 1938): 342–46, fig. 1, p. 344.

Part Two

Page 69: Schematic drawing adapted from Jean van der Tak, Carl Haub, and Elaine Murphy, "Our Population Predicament: A New Look," *Population Bulletin* 34, no. 5 (December 1979): fig. 1, p. 2.

Regarding number of years for population to increase by 1 billion:

Data for 1804 and 1927: United Nations Fund for Population Activities, *7 Billion Actions: The World at 7 Billion.*

Data for 1960–2011: United Nations, Department of Economic and Social Affairs, Population Division (2015). *World Population Prospects: The 2015 Revision,* DVD edition.

Data point for 8 billion in 2024 estimated from: United Nations (2015). Probabilistic Population Projection based on *World Population Prospects: The 2015 Revision*, Population Division, DESA. http://esa.un.org/unpd/wpp/

Page 71: Data before 1950: Population Division, Department of Economic and Social Affairs, "The World at Six Billion," 12 October 1999. Table 1, p. 5. Accessed at www.un.org/esa/population/publications/sixbillion/sixbilpart1.pdf. Data 1950–2015: United Nations, Department of Economic and Social Affairs, Population Division (2015). *World Population Prospects: The 2015 Revision,* DVD edition.

Page 73: Data 2015–2100: United Nations (2015). Probabilistic population projection based on *World Population Prospects: The 2015 Revision,* Population Division, DESA. http://esa.un.org/unpd/wpp/

Page 75: Data for 95% prediction interval 2015–2100: United Nations, Department of Economic and Social Affairs, Population Division (2015). *World Population Prospects: The 2015 Revision.* http://esa.un.org/wpp/Graphs/Probabilistic/POP/TOT/

Page 77: Adapted from Carl Haub and James Gribble, "The World at 7 Billion," *Population Bulletin* 66, no. 2 (July 2011): fig. 2, p. 3, and from chart at https://www.premedhq.com/demographic-transition

Pages 80–81: Crude birth rate: United Nations, Department of Economic and Social Affairs, Population Division (2015). *World Population Prospects: The 2015 Revision,* custom data acquired via website. http://esa.un.org/unpd/wpp/DataQuery/

Infant mortality: Ibid.

Life expectancy at birth: United Nations, Department of Economic and Social Affairs, Population Division (2015). *World Population Prospects: The 2015 Revision,* custom data acquired via website: http://esa.un.org/unpd/wpp/DataQuery/

Secondary school enrollment: World Bank. World DataBank: World Development Indicators. Series: Gross enrollment ratio, secondary, both sexes (%). Custom data acquired via website: http://databank. worldbank.org/data/reports.aspx?source=world-development-indicators

Female literacy rates: World Bank. World DataBank: World Development indicators. Series: Youth literacy rate, population 15–24 years, female (%). Custom data acquired by website: http://databank. worldbank.org/data/reports.aspx?source=world-development-indicators

Pages 82–83: Data for Uganda, Guatemala, India, and Germany from United Nations, Department of Economic and Social Affairs, Population Division (2015). *World Population Prospects: The 2015 Revision.* Custom data acquired by website: http://esa.un.org/unpd/wpp/DataQuery/

Pages 84–85: Figure adapted from chart created by Schuyler Null/ Wilson Center, *New Security Beat: The Blog of the Environmental Change and Security Program,* April 18, 2011. http://www. newsecuritybeat.org/2011/04/un-releases-early-results-of-global-population-projections/ (by permission of Schuyler Null)

Pages 86–87: Figure from Carl Haub, "In 2011, World Population Surpasses 7 Billion," Population Reference Bureau, October 2011, fig. 2. Data from United Nations, Population Division, *World Population Prospects: The 2010 Revision*, medium variant (2011). http://www.prb.org/Publications/Articles/2011/world-population-7billion.aspx

Page 89:

More developed regions comprise Europe, Northern America, Australia/New Zealand, and Japan.

Less developed regions comprise all regions of Africa, Asia (except Japan), Latin America, and the Caribbean plus Melanesia, Micronesia, and Polynesia.

Data points for 1900 from John D. Durand, "The Modern Expansion of World Population," *Proceedings of the American Philosophical Society* 11, no. 3 (June 1967): 136–59, table 1, p. 137.

Data 1950–2100 from United Nations, Department of Economic and Social Affairs, Population Division (2015). *World Population Prospects: The 2015 Revision*, custom data acquired via website: http://esa.un.org/unpd/wpp/

Page 91: See note for frontispiece.

Part Three

Page 105: Adapted from Jonas Salk, *The Survival of the Wisest* (New York: Harper & Row, 1972), fig. 9, p. 17.

Page 107: Ibid., fig. 10, p. 18.

Page 109: Ibid., fig. 11, p. 21.

Page 124: For a full discussion of the relationship between balanced wealth and human well-being, see Richard Wilkinson and Kate Pickett, *The Spirit Level: Why Greater Equality Makes Societies Stronger* (New York: Bloomsbury Press, 2009).

Page 129: Salk, *Survival of the Wisest*, fig. 21, p. 108.

Part Four

Page 149: Ibid., fig. 12, p. 23.

Part Five

Page 163: See note for frontispiece.

Page 176:

United Nations Interagency Group for Child Mortality. *Estimation, Levels and Trends in Child Mortality*—Report 2015.

World Bank, October 4, 2015. *World Bank Forecasts Global Poverty to Fall Below 10% for First Time.* http://www.worldbank.org/en/news/press-release/2015/10/04/world-bank-forecasts-global-poverty-to-fall-below-10-for-first-time-major-hurdles-remain-in-goal-to-end-poverty-by-2030

World Bank, *Overview.* Updated April 13, 2016. http://www.worldbank.org/en/topic/poverty/overview

World Bank, Graph: Poverty headcount ratio at $1.90 a day, 2011. http://data.worldbank.org/topic/poverty

Page 178: An exploration of this idea can be found in Jared Diamond's *The World until Yesterday: What Can We Learn from Traditional Societies?* (New York: Viking, 2012).

Related Readings

Brown, Lester. *Plan B 4.0: Mobilizing to Save Civilization.* Rev. ed. New York: W. W. Norton, 2009.

Diamond, Jared. *The World until Yesterday: What Can We Learn from Traditional Societies?* New York: Viking, 2012.

Eisenstein, Charles. *Sacred Economics: Money, Gift & Society in the Age of Transition.* Berkeley, CA: Evolver Editions, 2011.

Friedman, Thomas. *Thank You for Being Late: An Optimist's Guide to Thriving in the Age of Accelerations.* New York: Farrar, Straus & Giroux, 2016.

Harari, Yuval. *Sapiens: A Brief History of Humankind.* New York: HarperCollins, 2015.

Lane, Nick. *The Vital Question.* New York: W. W. Norton, 2015.

Lutz, Wolfgang, Warren C. Sanderson, and Sergei Schurboy, eds. *The End of World Population Growth in the 21st Century: New Challenges for Human Capital Formation & Sustainable Development.* London: Earthscan, 2004.

Mackey, John, and Rajendra Sisodia. *Conscious Capitalism.* Boston: Harvard Business Review Press, 2014.

Rifkin, Jeremy. *The Empathic Civilization: The Race to Global Consciousness in a World in Crisis.* New York: Penguin, 2009.

Sachs, Jeffrey. *The Age of Sustainable Development.* New York: Columbia University Press, 2015.

Wilkinson, Richard, and Kate Pickett. *The Spirit Level: Why Greater Equality Makes Societies Stronger.* New York: Bloomsbury, 2009.

Wilson, E. O. *Half-Earth: Our Planet's Fight for Life.* New York: W. W. Norton, 2016.

Wilson, E. O. *The Social Conquest of Earth.* New York: Liveright, 2012.

Photography Credits

We would like to thank all the photographers who contributed to this book. Their remarkable work has added dimension to its concepts and has grounded the work in the reality of our time.

Page 12: Hulton Archive / Stringer
Image of Jonas Salk used with permission of the family of Jonas Salk

Page 52: © Sebastian Liste / NOOR

Page 54: Nicolas Gompel

Page 56: Mediscan/Alamy Stock Photo

Page 58–59: Mediscan/Alamy Stock Photo

Page 60: © byrdyak / stock.adobe.com

Page 78: Alexandra Boulat / VII / Redux

Page 136: Associated Press

Page 164: © Pep Bonet / NOOR

Page 172: © Pep Bonet / NOOR

Page 190: Ron Haviv / VII / Redux

Page 192: Ron Haviv / VII / Redux

Page 194: Mattia Crocetti / LUZ /Redux

Page 196: Michael Zumstein / Agence VU / Redux

Page 198: © Pep Bonet / NOOR

Page 200: © Pep Bonet / NOOR

Page 202: Simone Donati / TerraProject / contr / Redux

Page 204: © Pep Bonet / NOOR

Page 206: Sim Chi Yin / VII / Redux

Page 208: © Sebastian Liste / NOOR

SPONSORS AND RESOURCES

NOOR is a collective uniting a select group of highly accomplished photojournalists and documentary storytellers focusing on contemporary global issues.

VII was created in 2001 by seven of the world's leading photojournalists and by 2005 VII was listed in third position in American Photo's "100 Most Important People in Photography." VII now represents 19 of the world's preeminent photojournalists whose careers span 35 years of world history.

THE DESIGN FUTURES COUNCIL is a DesignIntelligence network of design, product and construction leaders who explore global trends, challenges, and opportunities to advance innovation and shape the future of the design industries and the environment.

SON&SONS is a design and strategy agency that helps high potential companies and non-profits achieve transformational growth.

ACKNOWLEDGMENTS

The revision and redesign of this work have incorporated the very best aspects of Epoch B values, attitudes, and behavior. The spirit has been collaborative, generous, and creative, and all involved have been quite selfless in the pursuit of our goal of getting these ideas into the hands of those who will, we hope, carry us into a positive future.

David Dewane deserves all the credit for the inception of the republication. It was he who received a copy of *World Population and Human Values*, found that it spoke to him in a powerful way and contacted me to see if I would have an interest in reviving it. He went on to spearhead the entire enterprise. I have commented that, were this a film, he has functioned as an executive producer—sparking and nurturing the concept, identifying the need for and finding a wonderful and collaborative design firm, providing feedback at every stage of writing and production, helping to raise funds, bringing in collaborators as necessary, and moving us into crowdsourcing to allow the project to move forward.

Ted Eccleston was brought on by David in 2015 and has become an invaluable, creative, and essential member of the team. His intellect, enthusiasm and unwavering faith in the value of the book buoyed my spirits time and again and have made the final product what it is today.

We discovered Emily Gonzales because she had been using the ideas and images of the first edition for years in her consulting practice and in her life. She had such an understanding and such deep feeling for the essence of the book that we invited her to join us on. She has been a boundless source of energy, intelligence, and creative intuition over these last months, and we are thankful to have found her.

Courtney Garvin, about whom I cannot say enough, worked tirelessly, cheerfully, and creatively in being the lead designer on the project. One of David's goals in republication was to make something beautiful, and through Courtney's efforts we have something deeply beautiful and functionally designed.

Son&Sons Design, under the direction of Wade Thompson, contributed huge amounts of time and resources at extraordinarily low cost and has been an essential collaborator on the project. In particular, April Gardner and Christina Del Rocco were collaborators both creatively and organizationally.

Jim Cramer, chairman and founding principal of Greenway Group, cofounder and chairman of the Design Futures Council, and a friend of my father's is ultimately responsible for the republication; it was he who mentioned *World Population and Human Values* to David, which sparked this entire effort. His ongoing personal and professional support, as well as that of the Design Futures Council, has meant a great deal throughout.

Wellington Reiter of Arizona State University has also played an essential role. Without his help, encouragement, fund-raising, his genius for putting people together and getting to the heart of matters, I would not be writing these notes.

I want to name and thank our initial funders, Beverly Prior and Skidmore Owings & Merrill, who made it possible to undertake the main work of the redesign.

Sue Smalley has been both an inspiration and a vital resource. Her appreciation for my father's work, her spirit, and her excitement about the new edition coupled with helping us make key contacts took the project to a level we would not have achieved without her.

Arianna Huffington, who was friends with my father in the years he was working with these ideas, has been wonderfully supportive throughout our Kickstarter campaign and, more recently, in the publication of this volume. I thank her deeply.

I also want to express my appreciation to those who have provided us with such wonderful comments on the book: Phil Enquist, Lou Cozolino, Richard Wurman, Terry Tamminen, Ron Garan, Daniel Siegel, and Michael Mann.

The following have provided valuable input at various stages of the work: Daniel Schensul of the United Nations Population Fund, Hans Miller, Lou Cozolino, and Jody Becker.

Robbie Conal and Deb Ross hold a special place in my heart for many reasons, including being wonderful human beings and my best friends. No words can thank them enough for the life they bring to every moment.

I want to thank Ellen Scordato of Stonesong for her help and her belief in this project in our early search for a publisher. And my deep thanks to our wonderful copy editor, Karen Jacobson, for her careful work, support, and enthusiasm.

Elizabeth Carpenter played a huge role in bringing this book to press. Turned down, thoughtfully, by all our potential publishers, I was ready to give up. As a last attempt, I checked with Elizabeth. Generously, she personally took the book to Mike Shatzkin who in turn introduced me to David Wilk of City Point Press. He loved the book and took us on. I cannot thank her enough.

David Wilk has been incredible in his role as publisher and, importantly, in believing that the book could have a life in the broader market. In addition, I must acknowledge our marketing team of Steve Kroeter and Josh Schwartz as well as our publicist, Mary Bisbee-Beek. All three have been wonderfully dedicated to the project and remarkably effective.

Elizabeth Blackburn has provided the perfect foreword, touchingly understanding my father's deepest aspirations and values, and, in her characteristic way, gently but firmly explaining the importance of this slim volume. I am pleased to have found such a brilliant, articulate, and sensitive supporter to introduce the book to a broader audience.

My brothers, Peter and Darrell, have accompanied me not only through the completion of this book but also through the long and complex process of grieving for my father and of tying up the loose ends of his huge and complex life.

Françoise Gilot, my stepmother, who was there for the first edition, has been a wonderful champion of this new, colorful, and artful edition.

Ted Eccleston (and the entire team) thanks Cristel Zoebisch. She has been a source of support, constructive criticism, and inspiration every step of the way.

David Dewane (and again, the whole team) thanks his wife, Rachel, not just for her support and patience with David's generous dedication of time and creative energy to the project but for her sensitive, professional, and critical advice throughout. David also deeply thanks Andrew Balster, Pliny Fisk III, Tim Gentry, and Wendee Wolfson, all of whom were essential in getting him through to the finish.

I want to thank my wife, Elizabeth Shepherd, who has both the best eye and the biggest heart of anyone I know. She has quietly kept me on task with unflagging love and support and has been there with invaluable input on the design and text. She and our two sons, Ben and Hugh—both children of Epoch B in so many ways—have encouraged me throughout the project, believed in it, and made sure that I have seen it through to completion.

Last, I want to thank my father for these ideas, for his full and wonderful life, and for the legacy he left behind for all of our descendants.

SPECIAL THANKS

Special Thanks to our Kickstarter backers. Without their generosity and enthusiasm, the publication of this book would not have been possible.

Aaron, ABDOC Inc., Kyle Adair, Kyle Adams, Paola Aguirre, William Ahearn, Jun Ajima, Anthony Akins, Matt Alcock, Ghaleya Aldafiri, Fabian Alfvegren, John W. Algood, Noor 'Aliah, Stephanie Allen, Aaron Allon, Kathryn Alma-Nihte, Albara Alohali, Rodolfo Alvarez, Kunal Ambasana, Bharath Anandaram, Kathleen Anderson, Meadow Anderson, Stephen P. Anderson, Keith Andress, Peter Antal, Ayk Antil, Nikita Antonov, Michael H. Apffel, Matthias Arauco, Mim Armand, Stig-Jorund B. Arnesen, Robin Arnoff, Simon Arnold, James Atterbury, Nick Audette, Clement Auffray, Amy & Christian Austell, John Autin, Tal Azouri, Andrew Baartz, Peter Bachler, John A Baer, Lex Baer, Erin Baer, Andrew Balster, Rob Barnett, Kate Barrelle, Fulvio Bartolucci, Alexander Baumgardt, Jody Becker, Julie A. Behounek, Traci Belanger, Christophe Belina, Chris Beltran, Matti Berg, Tasha Rosow & Daniel Berger, Craighton Berman, Julio Bermudez, Melanie Bernstein, Allen Bey, Aman Bhageria, Sonya Bhavsar, Graham Bird, Janina Birtolo, Samir Biswas, Alan Black, Andrew Blaisdell, Andrew Boardman, Pamela Summit Bohn, Anna Boorstin, Paul M. Boos, Jonathan M. Borchardt, Matthew Borkowski, Clemi Boubli, Jenifer M. Bourcier, Jim Brady, Christoph Brandstatter, William Ryan Breen, Timothy Brewer, Jeff Bridges, Michelle Brinkos, Jayme Brisch, Duke Briscoe, Cathy Brown, Anthony Brzozowski, Daniel Bucherl, John Buchholz, Josias Buchweitz, Wayne Buente, Nathaniel Bullard, Travis Bunn, Sharon Burdett, Jeannie Burger, Ryan Burgess, Andy Bursavich, Maureen Busch, P. Buskirk, Peter Caesar, Kieran Calavan, Andrea Calderoni, Dimitri Callens, John Calzada, Edgar Camargo, Brian Canady, Amado Candelario Jr., Michael Ian Canepa, Paul Cao, Zachary Caplette, Marco Cardamone, Angie Cardenaz, Andre Cardoso, Britta Carlson, Maria Carrera, John

Caruso, John Cary, Jim Castrillon, Liz Cazan, Erkan Cetin, Vincent
Chang, Winston Chang, Kyle Chappell, Joseph Chin, Michael Chui,
Emmanuel Churchley, Maciej Cichocki, John Clemow, Garnett
Cohen, Mauro Marrs Coiro, Linda Colley, Nick Colombo, Brad
Columbine, Deb Ross & Robbie Conal, Conduit Sports, Cole Cooper,
Glenn Copeland, Cedric Cornez, Edward Cox, Willow Coyote, Jim
Cramer, Geoff Cranko, Sarah Craven, Jonathan Brett Crawley, Walter
F. Croft, Turil Cronburg, Nick Crosbie, Cody Cross, Ryan Culligan,
James Curcio, Anne Cutchin, Andrew Daley, Steve Daniels, Jed
Davies, Joshua Davies, Alice Davis, Jacqueline Day, Oria Jamar de
Bolsee, Vail De Capite, Tammy N. Dearie, Todd DeCarlo, Shari
Dehority, Scott Della Peruta, Karen Dellow, Dave Derycker, Neil
Deshpande, Frank Devine, David Dewane, James DeWitt, Axilleas
Diamantopoulos, Janel diBiccari, Howard Dickler, Christina Del
Rocco DiLegge, John S. Dimatos, Luke Dixon, Sandra Djuzic, Eugene
Dobry, Donna Dollner, Megan Leann King Dombeck, Julie Donohue,
Dotdash Dotdash, Annie Dourte, Botond Draskoczy, Michael
Hallback dSciarrone, Oliver Dudman, Charles Duffy, Lucas Dugrenot-
Felici, Mark Graham Dunn, Gerald Dupuis, Matt Eanes, Maximillian
Eberl, Bobby Eccleston, Johan Eckerstrom, William Edwards, Noha
El-Ghobashy, Brett Elmendorf, Graham Emonson, William Evanoff,
Jonas Evertsson, James Fackert, Graham Fagg, Dug Falby, Mia
Fantaci, Bill Farrell, Alexandre Ferre, Thomas Edward Fiello II,
Asaph Neiger Fijate, Todd Fischer, Tom Fisher, Ariel Fisk-Vittori,
Miles Fleming, Michael Flood, Ryan Flynn, Christian Foisy, Catherine
Maurice Foix, Andrew Foote, Duncan Fordyce, Michael Fornasiero,
Erin Foster, Maeve Fox, Cynthia Francis, Juliet Franco-Cooper, Mark
Frank, Bryan Frank, Dan Freeman, Jessica L. Freeman, Samuel B.
Freund, Jonathan Funkhouser, Luke Galambos, David Gallagher,
Emily Gallardo, Elizabeth Garcia, Jeff Gardiner, Steve Garguilo,
Thomas Gartenmann, Courtney Garvin, William Garvin, Einstein
Gautama, Richard G. Gelles, Matthew Genaze, Pier-Olivier Genest,
Ann Gerondelis, Shoshana Gerson Gerson, Nikolaus Gerteis,

Frank B. Gibney Jr., Ms. Coral Giffin, Eric Gilliland, Emmanuel Gilloz, Martin Giraldo, Anna Gladstone, Waco Glennon, Tobias Glueck, Eric Gobuty, David Godfrey, Charlotte Goethals, Eric Gold, Judith Roth Goldman, Emily A. Gonzales, Katey Goodall, Alina Gorina, Eric Gose, Sawyer Gosnell, Daniel Gould, Alberto Govela, Anne Graham, Jenn Graham, Adam V. Graham, Kiraah Grandberry, Donna Graves, John P. Gray, Laura Davis & Tjardus Greidanus Greidanus, Susan Grieder, Dave Griesbach, Coralie Grozner, Chris Guinnup, Rachel Gumpper, Greg Guthridge, Soraya Haas, Jonas Haefele, Sebastian Hagen, Jennifer Halbert, Stella Hall, Shannon Haltiwanger, Judith Roth Hanks, Simun Pauli Hansen, Mahan Harirsaz, Lindsay Harkema, Sara Ruckle Harms, Amy L. Harris, Branden Harvey, Sam Harvey, Max Haskvitz, Elizabeth Hass, Alia Hassan, Jeremy Hatter, Peter Haug, Ron Hazelton, Todd Hein, Wes Heinle, Paul Helinski, Meg Helm, Kendall Henderson, Robert Hendriksen, Miguel Hernandez, Eric Hester, Trevor Hetzer, Benjamin Heung, Joshua W. Hill, Erin Himrod, Liz Hincks, Caitlin Hirst, Trace Hobbs, Jan Hofkamp, Jess Holbrook, Katie Holbrook, Mark Holleman, Ronia Holmes, Richard Homann, Henry Hongmin Kim, Robert Luis Hoover, Alexis Hope, E.A. Hoppe, Hover Bike, Margaret Hovorka, Chris Hoyas, Randy Huebner, Jim Huffman, Erin Huizenga, Halit Firat Ilhan, Kevin Ingersoll, Timothy Inglis, Tummarong Ingpongpun, Niall Inverarity, Farhana Jabir, Brenden Jackson, Johnna Jackson, Charlotte Jacobs, David Jacobson, Param Jaggi, Steven Janssen, David Jaquet, Daniel Jaramillo, D'Arcy Jeffery, Sybren Jelles, Patricia Jenatsch, Sam Jenner, Vince Jennetta, Lianna Jewett, Johnathon, Valerie Johns, Gila Jones, Lionel Jones, Lloyd Joyce, Kerno Julien, Donna Kacmar, Melanie Kahl, Yasuhiro Kakegawa, Roman Kalantari, Jonathan Hise Kaldma, Kyle Kamka, Moniza Kanani, Meyer Kao, Eric Katz, Justin Kazmark, Guus Keder, Katherin Keech, Graeme Keleher, Patrick Gage Kelley, Steve Kelsey, Lauren Kennedy, Leon Kenyon & Cynthia Burlingham, Joanne E. Kerekes, Joel Kerner, Bradley Kerr, Jinny Khanduja, Constantin Kichinsky, Richard King, Robin Kissell,

Howard Kistler, Stefan Kjartansson, Jimmy Kjellstrom, Asher Klein, Derek Klinge, Kristin Klinkner, Carni Klirs, Lucia Knell, Jim Knoll, David J. Kochbeck, Scott Koehl, Willa Koerner, Jim Kohlmoos, Andrew Korf, David Korr, James Kovacs, Mark Kowatch, Bruce Kramer, Stanislav Krasovskyi, Eric Kretschmer, Scott Kroyer, Daniel Kubler, Rajiv Kumar, Shiva Kumar, Johan Kvarving Vik, Dylan Kwak, L.K. Labadi, Kurt Labuschewski, Jean-Noel Lam, Elisa Lamont, David Lang, Peter J. LaPrade, Chris Lauritzen, John Law, Chris Lawes, Larry S. Lazarus, Falk Lechtenborger, Cynthia Lee, Derek Lee, Eugene Lee, Gina Lee, Gregory Lee, Jacqueline Lee, Jono Lee, Miranda Lee, Philip Lee, Nathalie Leiba, Brittanie Leibold, Amber Lemons, Bastian Lengert, David Lenton, Sean Leow, Daniel Lepore, David Lesser, Patricia Lester, Kelvin Leung, Andrew Levin, Aran Lewis, Jeff Lewis, Kali Lewis, Steffen Linssen, Charles J. Lisle, John Little, Rodney Lloyd, Bryce Lord, Ross Losher, Ulrich Lott, Melissa Lucarelli, Ruben Lucendo, Nicholas Luiz, Michael Lujan, Edward Luna, Michael Lundquist, Chee Lup Wan, Kyle MacDonald, Paisley Mackie, Catherine Maddox, Hana Maguire, Bill Mahoney, Roni Malek, Elizabeth Elsas Mandel, Varun Mangla, Karolina Manko, Brandon Manning, Paul Manos, Paulette Manville, Ariane Matte Marchand, Alvise Marino, Beth Markley, Andrew Martin, James Martin, Jim Masland, Sam Mathis, Andrew Mayfield, Mary Helen Mays, Hilarie Mazur, Nikita McCauley, Sharon Ellen McClafferty, Greg McCormick, Matthew McCormick, Tom McGuinn III, Timothy McHugh, Kyle McKenna, Christine McKinnon, Kilian McMahon, Katey McTernan, Brent McWatters, AnaMaria Meca, Carlos Bernal Medina, Jose R. Mejia, Samuel Melancon, Mell, David Meltzer, Merlion, David Meshoulam, Michelle Michelle, Fabian Middelmann, Stephanie Mikulich, Sebastian Milanes, Erin Miller, Karl Miller, Kirby Miller, Patrick Miller, Lucas Milone, Jan Minium, Yessenia Miranda, Miguel Nunes de Miranda, Stephen Mitton, Drew Mohoric, Phil Molloy, John Monguillot, Abby Monroe, Katharine Moore, Skylar Moran, Ruben Morgado, Robert Morgan, Tim Morgan, Mariano Moro, Chad Morris, Chris Motley, Charlena Moy, Don Moyer, Wesley Mudge, Kathleen

Muller, Mary Mullins, Colin Murray, Jaclyn Nagy, Zev Nathan, Bryan Nay, Kyle Neath, Alexandru Nedel, Katrina Neff, Jon Neseth, Randi Michelle Neville, Cal Newport, Jake Nickell, Jonathan Nienaber, Juan Nieto, Michael Niosi, Michael Nock, Lindsay Ann North, Imanzah Nurhidayat, Kevin O'Donnell, Mark Ohlinger, Bob Olodort, Kristian Olsen, Thomas Malmqvist Olsson, Joanne M. Ong, Deirdre Opp, Rodrigo Ortiz Vinholo, Paulina Ospina, Adrienne Ou, Lisa Overton, Frantisek Pac, Michael Packer, Michael Page, Alfie Palao, Regina Pally, Joseph Palombo, Davide Papotti, Phil Parcellano, Mel Parks, John Parnitzke, Abner Parzen, Chase Pashkowich, Vaclav Paur, Michelle Pawson, Jonathan PayCheck, Hamilton Peek, Stephanie Pereira, Luis Perez, Oscar Perez, Jorge Oracio Perez Prieto, Paolo Perez-Rubio, Allan Pernot, Clarke Peterson, Vlad Petrea, Connie Petska, Andre Piazza, Dominique Pickett, Alex Pierer, David Pigott, Stefano Piroddi, Tom Plaskon, Genevieve Poist, Emma Poppante, Krishna Prasad, Greta Prozesky, Zachary Prusak, Colleen Pundyk, Eugene Quah Ter-Neng, Abdul Qutub, Tibor Racz, Kartheesan Ragavan, Nino Rajic, Luke Raymond, Arturo J. Real, Timothy Reese, Bobby Reichle, Ray Reigadas, Duke Reiter, Arlen Rencher, Rob Reul, Raul Reymundi, Francine Wolf Rickenbach, Anna Rickert, Elissa Rinehart, J.T. Ripley, Mikey Riva, Douglas Roberts, Evan Robertson, Nichole Robertson, T. Robeyns, Srebrenka Robic, Graham Robson, Matthew Roche, Moacir Rodrigues, Abner Rodrigues Neto, Ben Roffer, Paul Roger, Samuel Rokebrand, Michal Rosenn, Genevieve Rosow, Harry Ross, Mary Ross, Steven Ross, Mark Rossen, Daniel Rossi, Noah Rowlett, Troy Rubingh, Benjamin Russo, Ben Rybolt, Thomas Rybolt, Miguel Sa, Siddharath Saigal, Azmir Saliefendic, Darrell Salk, Jesse Salk, Peter Salk, Shane Salk, Donald Salter, Pornsatit Samattiyadeekul, Roque Sanchez, Seraluna Sanchez, April Sandefer, Eric Damon Sanderson, George Sarantopoulos, Todd Sarty, Andrew Scaife, Markus Schaefer, Scott D. Schaffer, Andreas Schanzenbach, Diane Schear, Fredrik Schiren, Tim Schmitt, Rob Schnabel, Kyle Schneider, Josh Schoenwald, Benjamin Schone, Brian Schottlaender, John Schretlen, Matthew Schultz, Nate Scott, Jeffrey

Scott Milam, Linda Sepe, Vereshchako Sergey, William Sergison, Serio Collective, Gopakumar Sethuraman, Jonathan Sexton, Joe Shaleen, Aleksandra Shander, Sanford Shapiro, Debrorah Shaw, Timothy M. Shead, Evan Sheehan, Stephen Shell, Sam Shepherd, Victoria Shepherd, Kel Sheppey, Sherman, Joseph Shields, Tammy Shirley, Rebecca Sibley, David & Jo-Nell Sieren, Anthony Simeone, Jessica Simmons, Johnny Simmons, Derek Simpson, David Smigielski, Justin Smith, Daniel Smith, Sebastian Solander-Bruus, Roberta Solit, Juhan Sonin, Andre Sorensen, Renier Soto, Jan Soucek, Thomas Sowell, Victoria Sparks, Frank B. Spencer, Kent Spillner, Angela Spinazze, Alexander Spoor, Karen Spratley, Austin Stahl, Jeff Stanton, Scott Steele, Daniel Stein, Ken Shubin Stein, Ron Stelmarski, Anna-Lucia Stone, Paul Storer, Amy Danforth Stover, Nick Stoynoff, Jeffrey Strawbridge, Yancey Strickler, Robert Strohsahl, Ben Strother, Simon Stroud, Thomas A. Suberman, Jan-Erik Suttle, Joseph Sweeney, Stephen Swicegood, Gittit Szwarc, Harrison Tan, Venetia Tay-Mozilla, Barbara Taylor, Charlie Taylor, Dana Taylor, Gail Taylor, Tenzubg Thabkhe, Hung Thai, Montagne Thibaut, Sean Thiers, Jeff Thompson, Wade Thompson, Runar Thorstensen, Sjors Timmer, Robbie Tingey, Satyam Tiwary, Emi Tomijima, Toby Tomkins, Darren Torpey, Connor Towles, Patrick Ryanb Triato, Kate Trimble, Konstantinos Tsetsonis, Richard Tucker, Tracey Tufnail, Michelle Turcotte, Chris Urban, Lyndon Valicenti, Matthew Vallevand, Jeroen van Dam, Jason Van Nimwegen, Daniel Veale, Jennifer Vickerman, Juan Pablo Vidal Venegas, Tommi Virta, Gail Vittori & Pliny Fisk III, Mark Vlasic, Alexander von Schlinke, Florian von Teppner, John Waclawski, Ann Wagstaff, Andrew Walsh, Eric Damon Walters, Lawrence Wangh, Claudine Ward, Alyssa Warner, Denise P. Watson, Bill Webster, Jan Welke, Kristen Weller, Kai Wenas, Shannon Wendt, Doug Wessel, Janet Whatmough, Caleb Whitaker, Dave Whitling, Pascal Wicht, Stephani Widmer, Lyle Wiedeman, Nicholas Wiegand, Jed Wilcox, Cedric Wilhelm, Blake Wilkinson, Kristin Will, Rick Williams, Nicholas Wilson, Stian

Wilson, Jennifer Wilson, Melanie Winter,
Jeff Wismann, Alexander M. Wolf, Wendee Wolfson, Peter Wai-Chor
Wong, Kevin Wong, Maggy Wong, Ming Woo, Michael Wood, Larry
C. Woods, Jeff Woolf, Ben Wooliscroft, Pete Woshner, James Wright,
Jenny Wu, Eddie Xing, Minfei Xu, Yimeng Xu, Jeff Yamauchi, Lily
Yang, Mykel Yee, Jason Yegge, Kevin Yeh, Fernando A. Yepez, SiCong
Yi, Kaloyan Yordanov, Robert Young, LinYee Yuan, Tieg Zaharia,
Saarim Zaman, Michael Zang, Fabian Zeidler, Dan Zimmerman,
Craig Zimmers, Oliver Zoellner, Dimitrios Zoulis

This book is printed for City Point Press by Friesens Corporation, Altona, Manitoba, Canada. The paper is Finch Opaque Vellum. The text is composed in Surveyor and Aaux Next.

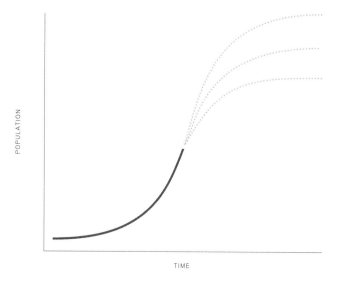

POPULATION

TIME

The course of epochal change is not predetermined.
It is subject to our influence.